DATE DUE

A CONFIGURATIONAL
MODEL OF MATTER

STUDIES IN SOVIET SCIENCE

1973

Motile Muscle and Cell Models
 N. I. Arronet

Densification of Metal Powders during Sintering
 V. A. Ivensen

The Transuranium Elements
 V. I. Goldanskii and S. M. Polikanov

Pathological Effects of Radio Waves
 M. S. Tolgskaya and Z. V. Gordon

Gas-Chromatographic Analysis of Trace Impurities
 V. G. Berezkin and V. S. Tatarinskii

A Configurational Model of Matter
 G. V. Samsonov, I. F. Pryadko, and L. F. Pryadko

Complex Thermodynamic Systems
 V. V. Sychev

Central Regulation of the Pituitary-Adrenal Complex
 E. V. Naumenko

Crystallization Processes under Hydrothermal Conditions
 A. N. Lobachev

STUDIES IN SOVIET SCIENCE

A CONFIGURATIONAL MODEL OF MATTER

G. V. Samsonov
Institute of Problems in Materials Science
Academy of Sciences of the Ukrainian SSR
Kiev, USSR

I. F. Pryadko
A. M. Gor'kii Kiev Pedagogical Institute
Kiev, USSR

and

L. F. Pryadko
Institute of Problems in Materials Science
Academy of Sciences of the Ukrainian SSR
Kiev, USSR

Translated from Russian by
Albin Tybulewicz
Editor: Soviet Physics—Semiconductors

CONSULTANTS BUREAU • NEW YORK - LONDON

Library of Congress Cataloging in Publication Data

Samsonov, Grigoriĭ Valentinovich.
  A configurational model of matter.

  (Studies in Soviet science)
  Translation of Konfiguratŝionnaĭa model' veshchestva.
  Bibliography: p.
  1. Free electron theory of metals. 2. Valence (Theoretical chemistry) 3. Configuration
space. 4. Transition metals. 5. Transition metal alloys. I. Priadko, Ivan Fedorovich, joint
author. II. Priadko, Leonid Fedorovich, joint author. III. Title. IV. Series.
  QC176.8.E4S2413   530.4'1   73-83893
  ISBN 0-306-10890-9

Grigorii Valentinovich Samsonov was born in 1918 near Leningrad. He was graduated in
1940 from the M. V. Lomonosov Institute of Fine Chemical Technology in Moscow, where
he specialized in the fine chemical technology of inorganic substances. In 1958 he was
awarded the degree of Doctor of Technical Sciences and in 1959 he became a professor.
At present he heads the Laboratory of Refractory Compounds and Rare Metals at the Institute of Problems in Materials Science of the Academy of Sciences of the Ukrainian SSR,
and holds the Chair of Powder Metallurgy at the Kiev Polytechnic Institute.

Ivan Fedorovich Pryadko was born in 1940. In 1963 he was graduated from the Radio-
physics Department of Kiev State University. He has lectured at the Kiev Polytechnic
Institute and worked at the Institute of Problems in Materials Science of the Academy of
Sciences of the Ukrainian SSR. In 1968 he was awarded the degree of Candidate of
Technical Sciences and at present he is lecturer at the Kiev Pedagogical Institute.

Leonid Fedorovich Pryadko was born in 1944. In 1968 he was graduated from the Radio-
physics Department of Kiev State University. In 1972 he was awarded the degree of Candidate of Physico-Mathematical Sciences. He is currently working at the Institute of Problems in Materials Science of the Academy of Sciences of the Ukrainian SSR.

The original Russian text, published by Naukova Dumka in Kiev in 1971, has been corrected by the authors for the present edition. This translation is published under an
agreement with Mezhdunarodnaya Kniga, the Soviet book export agency.

КОНФИГУРАЦИОННАЯ МОДЕЛЬ ВЕЩЕСТВА
Г. В. Самсонов, И. Ф. Прядко, Л. Ф. Прядко

KONFIGURATSIONNAYA MODEL' VESHCHESTVA
G. V. Samsonov, I. F. Pryadko, and L. F. Pryadko

© 1973 Consultants Bureau, New York
A Division of Plenum Publishing Corporation
227 West 17th Street, New York, N. Y. 10011

United Kingdom edition published by Consultants Bureau, London
A Division of Plenum Publishing Company, Ltd.
Davis House (4th Floor), 8 Scrubs Lane, Harlesden, London, NW10 6SE, England

Printed in the United States of America

Preface to the American Edition

The book we are presenting to American and other English-speaking readers is a review of the work on the electron structure of elements, alloys, and compounds, which was started back in the fifties. This work gradually grew into a system of ideas on the electron structure of condensed matter which is now known as the configurational model.

This model is based on the assumption of the preferential formation of the most stable configurations of the localized valence electrons in condensed matter. The existence of these stable configurations and the exchange of electrons with the delocalized (collective-state) subsystem determines those properties which are related to the electron structure. The conclusions which can be drawn from the applications of the configurational model are only qualitative but they explain quite clearly the nature of various properties of condensed matter, and they are helpful in the search for materials with specified properties.

The American edition has been corrected and supplemented in many minor respects. Moreover, the opportunity was taken to revise thoroughly the section on the fundamentals of the configurational model in the light of the latest theoretical developments. Other parts of the book have been shortened to eliminate material which is not of fundamental significance or has not yet been developed sufficiently fully.

The authors are grateful to Plenum Press for the opportunity to make these revisions before the publication of the American edition and thus provide readers with the latest status report on the configurational model. The authors hope that this book will be useful in the search for new materials in various branches of technology and in the development of the theoretical basis of the electron structure of condensed matter. An attempt has been made to present this theoretical foundation in such a way that it will be useful to the widest possible circle of materials scientists.

The authors will be glad to receive any comments which will help in the revision of later editions.

G. V. Samsonov
I. F. Pryadko
L. F. Pryadko

Preface

The rapid development of the physics and chemistry of solids has made it easier to find the materials required in nuclear power engineering, semiconductor and insulator technology, the direct conversion of thermal energy into electrical power, and space engineering. Great progress has been made in the development of improved protective and wear-resistant coatings for machine parts and mechanisms.

In spite of these successes, the search for new materials remains largely empirical. Therefore, various models of the condensed state of matter have been developed, which — while they do not provide the final answer — do make it easier to search for new materials. Moreover, they ensure that the approach to materials science problems is more deliberate and purposeful. From the theoretical point of view, these models are temporary "constructs" which will be replaced by more refined models as the theory of solids and liquids is developed. The models used so far can be described in the words of R. P. Feynman: "This game of roughly guessing at a family of relationships is typical of the first encounters with nature preceding the discovery of some deeply significant and very important law." Models of this kind include, for example, the density-of-states model and the configurational model of condensed matter. The latter model is discussed in the present book. It originates from certain intuitive ideas in chemistry and physics. For example, back in 1933 L. V. Pisarzhevskii postulated that regrouping is possible between the s and d electrons in transition metals and that such regrouping can give rise to new electron configurations corresponding to "passive" isomers.

More recently, similar ideas have been applied to catalysts, and it has been concluded that the minima of catalytic activity are usually encountered for systems with the d^0, d^5, and d^{10} cation configurations, whereas the maxima correspond to configurations intermediate between these three states.

H. A. Bethe states in his "Quantum Mechanics" that "The binding energy increases until a shell in a transition metal acquires five electrons." Again he says: "...the order in which all shells are filled is governed by the Pauli principle and by the condition of m i n i m u m   e n e r g y   o f   t h e   r e s u l t a n t   e l e c - t r o n   s t a t e s" (our emphasis). Bethe also says: "Thus, chromium has, in place of a 4s electron, an additional 3d electron, which gives the configuration $4s3d^5$. This means that the half-filled shell is   s t a b l e.   In manganese we encounter a $4s^2$ configuration in addition to the stable $3d^5$ state; the $4s^2$ configuration is retained right up to nickel. Next, the   a d d i t i o n a l   s t a - b i l i t y   of the $3d^{10}$ shell of copper again results in one electron being left in the 4s state." The letter-spaced concepts are of basic importance in the configurational model and they also define the terms used in this model, particularly the stable configuration.

Chemists have long been aware of the stability of the electron configurations of the atoms of helium (s^2) and the other inert-gas atoms (s^2p^6), and of the quasistable sp^3 configuration of the carbon-atom electrons in diamond. In physics, we encounter the stable Cooper pairs of electrons.

The ideas of configurational localization have grown in various countries in different forms and on different theoretical bases. Physicists have studied many-electron systems, bearing in mind the specific interactions which determine the electron structure. An allowance for such interactions is particularly important in view of the difficulties encountered in the use of one-electron theories in the interpretation of x-ray diffraction, neutron diffraction, and x-ray spectroscopic data, and of the results of studies of electrical and structure parameters. Many phenomena cannot be explained at all without allowance for these interactions. This applies, for example, to polymorphism and the relationships that govern it; to the influence of admixtures on the properties and structure of alloys; to the behavior of solids at high pressures; to the nature of solid-liquid phase transitions; to surface phenomena; and to emission properties.

The configurational model of matter may prove of importance in biochemistry and biophysics. This model can be used quite easily by biologists and it has already been applied in studies of the toxic properties of elements and compounds.

The present book gives the theoretical fundamentals of the configurational model and discusses the applications of this model in the interpretation of the physical and chemical properties of transition metals and their compounds with nonmetals. These materials are of great importance because of their high-temperature applications.

The authors will be grateful for critical comments and they hope that, in spite of its unavoidable deficiencies, the book will be useful to a wide circle of specialists developing new materials and using them in solving the problems encountered in modern technology.

Contents

Introduction

The problem of the development of new materials with specified properties can be approached in several basically different ways. One is the analysis of the accumulated experimental data and the derivation of the relationships between various properties from these data. A different approach consists of the extrapolation of the properties of well-known materials to those which have not yet been studied but which belong to the same class or type and have identical or similar crystal structures. These two ways and similar ones are based on the known experimental information and involve a "reorganization" of such information without answering the question as to why a particular material has these and not other properties, why it is a refractory, why it has this and not some other specific heat, and generally, why it behaves in this way and not another. Obviously, answers to all these questions can be obtained only if we know the functional relationships between the electron structure and the properties of the material. Such relationships can be deduced from suitable concepts relating to the structure of matter, which must be based on the electron structure of atoms and which demonstrate the direct relationship between this structure and the macroscopic properties.

The most direct method for the determination of the electron structure of matter is the solution of the Schrödinger equation for a many-electron system with a suitable allowance for the largest possible number of interactions and a subsequent comparison of the results obtained with those deduced from experimental studies of the electron structure by methods such as x-ray, neutron, or electron diffraction, magnetic resonance, the Mössbauer effect, and positron annihilation. Unfortunately, the fundamental

1

information obtained in this way does not fulfil the requirements
of modern materials science. Therefore, it is often necessary to
use a different approach, in which models are developed from the
data on the physicochemical properties of various materials.

The application of the Heitler—London model (also known as
the Heitler—London—Heisenberg model) and the Bloch band
theory (collective-state model) to the transition metals and their
compounds has not been very successful in providing explanations
of their properties. It has been found that some properties of
these materials can be understood only on the basis of the theory
of collective-state electrons, whereas others can be understood
only on the basis of the model of localized Heitler—London states.
The band theory method is often preferred because the mathe-
matics of the Bloch approach is the most developed. However, in
dealing with chemical problems, the Heitler—London valence bond
method gives better results than the Bloch collective-state (molecular orbital) model. The Bloch and the Heitler—London approaches are, in a sense, mutually exclusive. In the theory which
considers localized states, the electron configurations are filled
in accordance with the Hund rules, whereas in the band theory,
the electron configuration which distinguishes one element from
another is lost completely and the configuration of a given atom
is built up in accordance with the lowest-spin rule, i.e., in an
anti-Hund manner. It has recently been established that the dual
nature of electrons in solids with narrow energy bands is due to
the fact that the intraatomic and interatomic interactions are
comparable. The influence of the intraatomic correlation on the
properties of a substance has been demonstrated in the theory of
thermionic emission [2], in the formation of crystal structures of
the transition metals [3], in electrotransport phenomena [4], and
in discussions of the nature of magnetism [5-7]. A rigorous quantitative approach to the configurational model can offer a combination of the more successful features of the Heitler—London
and the Bloch limiting models, and it can bridge the gap between
these two models, i.e., between the molecular orbital and the
valence bond methods or, in other words, between the special
properties of a given element, which are important mainly in
chemistry, and the band nature of the electron-energy spectra,
which is usually the starting point in many physics theories. In
the configurational model, the band nature of the energy spectra

is allowed for by postulating the existence of fluctuations of the electron configurations in a solid and the intraatomic correlation effects are taken into account more fully than in the band theory.

CHAPTER I

Configurational Model of Solids

1. Fundamentals of the Model

Two theories of the electron structure of crystals, the band theory of Bloch and the Heitler – London theory (also known as the Heitler – London – Heisenberg model), provide satisfactory explanations of most of the relationships governing the properties of solids with relatively simple electron structures [8-32]. However, these theories fail to account properly for the experimentally determined properties of d-type transition metals, whose atoms have partly filled inner shells in the free state, and of compounds of these metals. This is due to the complex electron structure of the substances with heterodesmic bonding. Therefore, it is necessary to review critically the assumptions and the approximations on which these simple theories are based. Since the effects which are not explained by these theories are important also in the limiting cases of alkali metals or dielectrics [33-35], we may conclude that these effects are important in the electron structure of all crystalline substances.

It is difficult to make a consistent allowance for the interactions which occur in condensed matter. However, an analysis of the physicochemical properties of the transition metals, their alloys, and compounds shows that the most important factors which often determine their behavior under various conditions is the degree of localization of the electrons and the stability of the valence shells in free atoms which are condensed to form a solid.

The ideas developed in the present monograph are based on the following principal assumptions [36-39]:

1. When free atoms are condensed to form a
solid or a liquid, the valence electrons of these
atoms split into localized and collective-state
groups.

The concept of localized electrons was first used in the quantum theory of dielectrics. It was later found that many properties of metals could also be explained by the localization of electrons. The widespread assumption that the band theory of Bloch is more suitable for the description of the properties of metals [40] has been criticized in connection with the interpretation of the magnetic moments of the transition metals [41]. The localization (at least partial) of the d electrons is supported by the existence of spin waves and by the Langevin nature of paramagnetism above the Curie or Néel points of some ferromagnets or antiferromagnets. The theory of the magnetic properties and the thermodynamics of magnetism developed on the basis of the localized Heitler–London–Heisenberg model have been highly successful in explaining and predicting the temperature dependences of the magnetic properties. Most of the mechanisms of the exchange interaction in ferromagnets are also associated with the existence of localized magnetic moments.

The bonding in the transition metals belonging to the first half of each of the long periods is heterodesmic [42]. The bonding in these materials has a strong metallic component as well as a considerable covalent component, which is easiest to interpret by the concept of localized electrons obeying the Hund rules, in contrast to collective-state electrons whose spins are paired at each lattice site.

The idea of electron localization has recently been confirmed by the experimental results derived from x-ray diffraction analyses of the transition metals [43-49], from electrotransport data [50], and from the Hall effect [51].

A theory of the localization of electrons in metals has been developed only quite recently in spite of the fact that Slater pointed out a long time ago [52] that a correct allowance for the electron-electron interaction can explain the localization even within the framework of the band model. Friedel, Leman, and Olszewski [53] extended the band theory to cover the case of localized electrons and demonstrated that the tendency for the localization of spin moments should be strongest for metals with half-filled bands. Other workers [54-56] used the band theory to explain the Bragg reflection and the inelastic scattering in the paramagnetic state above the Curie point. The correlation effects can ensure the validity of the Heisenberg exchange Hamiltonian in the case of metals [57] and can give rise to metal−semiconductor transitions in complex substances [50].

2. The localized fraction of the valence electrons in a solid forms a spectrum of configurations.

Lomer and Marshall [58] used the direct experimental data (obtained by x-ray diffraction) on the localization of the valence electrons in the transition metals [43] (refined later in [14]). However, Lomer and Marshall put forward no hypothesis relating to the origin of localized electrons. They suggested that such electrons are responsible for the magnetic properties of the transition metals. The localization of electrons was later used in many models. The internal structure of the localization is discussed in [29, 59, 60] on the assumption that electrons can be divided sharply into two subsystems e_g and t_{2g} by the internal lattice field. The first of these subsystems can be regarded as localized and the second as responsible for the collective-state effects in metals. However, according to Herring "...it is very doubtful that the crystal-field anisotropy is large enough to make a clear separation of this sort" (p. 5S in [28]). A similar opinion was voiced by Brooks [40]. The subsequent quantitative calculations [61, 62] did indeed show that the crystal-field effects in pure metals are negligible compared with the effects associated with the structure of the d energy band [20].

Van Vleck [63] pointed out that the d electrons could be described by the configurational model. This approach provides a satisfactory description of the magnetic properties of nickel and explains well the correlational interaction in condensed systems. The "jump" time of an electron from one site to another is governed by the width of the d band: it is less than 10^{-13} sec and, therefore, configurational fluctuations cannot be detected experimentally by diffraction methods.

The Van Vleck approach to the energy spectra was justified by Hubbard [64] who showed that the correlational interaction in

narrow energy bands forces electrons to spend most of their time
near atomic sites and to jump from one site to another as a result of the overlap and of translational symmetry in a crystal.
In this model, the density of the electrons in the spaces between
atoms is assumed to be low, which is just another way of saying
that the d bands of the transition metals are narrow. This means
that the motion of electrons in a many-electron system can be
described correctly with the aid of the Wannier functions, which
are similar to the wave functions of the electrons in a free atom.
In most of those cases when the detailed structure of the d bands
is unimportant, we can exploit the fact that the gaps between the
energy levels of the configurations with different numbers of electrons are considerably larger than the multiplet splitting between
the levels with the same numbers of electrons. If we neglect the
multiplet splitting, we easily arrive at the configurational description of a system of interacting electrons in a crystal, which
reduces the problem of a solid to that of atomic spectroscopy and
to a comparative estimate of the stability of the various configurations. This conclusion is in accord with the idea of Slater [52]
and Zener [26] that the interatomic interactions govern, to a considerable degree, the properties of solids. This makes it possible
to use the connection between the relationships governing free
atoms and the relationships governing condensed media, i.e., to
establish the genetic relationship between elements in the free
and condensed states.

3. The most stable configurations in a given
spectrum can be empty, half-filled, or completely filled.

The relative stability of the various configurations of free
atoms can be established by considering the periodic dependences
of the first five ionization potentials on the atomic number (Fig. 1).
The sharpest maxima in the dependences of these potentials correspond to completely filled inner s^2p^6 shells of the inert gases.
The binding energy between electrons and nuclei is also found to
increase for elements with the s^2 or d^{10} valence shells. Such
maxima can be seen clearly for zinc, cadmium, and mercury,
which have the $d^{10}s^2$ configurations, and for copper, silver, and
gold ions. Maxima are also observed for the half-filled p shells
of the atoms of N, P, As, Sb, and Bi, and those of O^+, S^+, Se^+,
and Te^+ ions. A considerable stability is exhibited by the half-

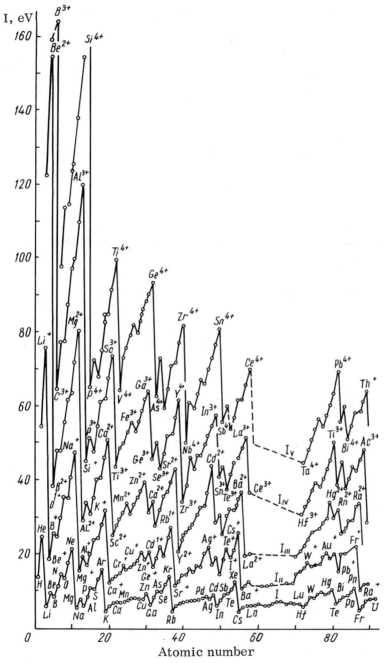

Fig. 1. Dependences of the first five ionization potentials on the atomic number.

filled shells d^5 (chromium, molybdenum, tungsten) and f^7 (gadolinium), as pointed out in [12].

The investigations carried out by Bondarev [65] have demonstrated convincingly the stability of the d^0 configurations. The stability of some of the configurations mentioned above has been used in the ligand field theory [66]. The modern theory of catalysis shows that the energy of stabilization by the crystal field vanishes when the number of electrons in the d shell is 0, 5, or 10 [67]. An analysis of the properties of rare-earth metals has indicated that the f^0, f^7, and f^{14} configurations are also energetically stable.

According to Pauling [68, 69], the strongest covalence of the bonding should be observed in the middle of each period. The hypothesis of the exceptional stability of the half-filled and filled configurations in solids was used by Hume-Rothery, Irving, and Williams [70], who refined the Pauling model but did not attempt to provide a unified theory. This attempt was made by Engel [71-77] and Brewer [78-83], whose views were recently analyzed by Hume-Rothery [84]. The model advanced by these workers uses the idea of configurational stability in a consistent manner and explains the observed variations of the binding energy of various elements in the periodic table. The Brewer-Engel model is very similar to that developed in the present book but is basically a variant of the theory of valence bonds, which limits the range of its validity in the explanation of the properties associated with the partial transfer of electrons to the collective state.

Thus the accumulated experimental data and the theoretical investigations indicate that the most stable configurations are s^2, sp^3,† s^2p^6, d^0, d^{10}, f^0, f^7, f^{14}.

4. The energy stability of the electron configurations is a function of the principal quantum number of the valence electrons from which these configurations are formed.

According to the model under consideration, an increase in the principal quantum number of the valence electrons results

---

†The sp^3 configuration is a hybrid of the stable s and p^3 configurations and is formed from s^2p^2 by one-electron s−p transitions. The sp^3 configuration is favored by the energy considerations [85, 80], according to which the relative values of the binding energies are 1.93, 1.99, and 2.0 for the sp, sp^2, and sp^3 configurations, respectively.

TABLE 1. Binding Energies
(per bonding electron)

Principal quantum number	E_{sp}, kJ/mole	E_d, kJ/mole
1	335	—
2	167	—
3	84	109
4	67	126
5	63	151

in an increase in the stability of the d^n and f^n configurations in a given group, and in a reduction of the stability of the $s^x p^y$ and s^2 configurations. This is supported by the nature of the dependences of the ionization potentials of free atoms on the atomic number. It is evident from Fig. 1 that the ionization potentials of the d valence electrons increase and those of the sp electrons decrease when the principal quantum number becomes larger. This applies also to configurations in solids, as found by Engel [77] from the dependence of the binding energy (per bonding electron) on the orbital and principal quantum numbers of the electrons (Table 1).

Quantitative calculations of the stability of the various d^n configurations and of their statistical weights can be carried out only within the framework of the many-electron approach with a sufficiently rigorous allowance for the correlation effects, which are particularly important in the case of the d- and f-type metals. Nevertheless, some general relationships governing the configurational spectra can be deduced by a purely qualitative approach, in which the probabilities of the formation of different configurations are considered [87-90]. Clearly, when the electron configuration of a free atom approaches one of the stable states, the statistical weight of this state in the transition metal increases. This increase should be observed also for crystals of metals with high values of the principal quantum number of the valence electrons. For example, in the case of scandium, a free atom contains only one d electron and the probability of the appearance of the d^0 and d^1 states is extremely high. As we approach the half-filled state, the probability of the formation of the more stable d^5 states increases. The highest statistical weight of these states is encountered in metals whose free atoms have the d^5 configuration (this configuration is disturbed somewhat because of the need to transfer some of the d electrons to form bonds).

The statistical weight of the stable d^{10} states is high in metals located at the ends of the long periods.

A consistent quantum-mechanical approach to the configurational model is based on the Hamiltonian of the electron system, which can be divided into delocalized (collective-state) and partly localized electrons:

$$H = H_s + H_d + H_{sd} \tag{1}$$

where H_s is that component of the energy operator which describes the delocalized (collective-state) electrons; H_d is the Hamiltonian of the d electrons; H_{sd} represents the hybridization of the states.

The collective-state effects can be included in the configurational model because the delocalized electrons are described by the Bloch wave function

$$\psi_k = \frac{1}{\sqrt{v}} U_k(r)\, e^{ikr} \tag{2}$$

with a factor $U_k(r)$ which is nearly constant, and because these electrons have the anti-Hund distribution. The density of these electrons is largely governed by the number of vacancies in the d shells and by the tendency for stable configurations to form in solids. At low densities, the delocalized electrons contribute to the bonding but at high densities they weaken the lattice.

The second term in Eq. (1) should be expressed in terms of the approximation that gives the many-electron effective Hamiltonian of a solid:

$$H_d = H_0 + H_{corr} + H_{exc}, \tag{3}$$

where H_0 is the one-electron component of the energy operator; H_{corr} represents the correlational interaction between electrons; and H_{exc} represents the exchange interaction. In narrow d bands, the second and the third terms of the above equation are dominated by the interactions occurring at a single lattice site. If the Coulomb interaction at one site is diagonalized, the second term of Eq. (1) can be represented in the form [8]

$$H_d = \sum_{m,\Lambda} \epsilon^\Lambda Z_m^\Lambda + \sum_{\substack{m,n \\ \Lambda_1,\Lambda_2,\Lambda_3,\Lambda_4}} B_{mn}^{\Lambda_1\Lambda_2\Lambda_3\Lambda_4} Z_m^{\Lambda_1\Lambda_2} Z_n^{\Lambda_3\Lambda_4}. \tag{4}$$

Here, Z^Λ represents the statistical weight of the atoms in the
state $|\Lambda)$, which is defined by the quantum numbers L and S; ε^Λ
is the energy parameter corresponding to this state; $Z_m^{\Lambda_1 \Lambda_2}$ is
the operator which represents the change in the electron con-
figuration at the site m from the state $|\Lambda_1)$ to the state $|\Lambda_2)$;
$B_{mn}^{\Lambda_1 \Lambda_2 \Lambda_3 \Lambda_4}$ is the matrix element of the transition indicated in
Eq. (4). The quantities $B_{mn}^{\Lambda_1 \Lambda_2 \Lambda_3 \Lambda_4}$ and $Z_m^{\Lambda_1 \Lambda_2}$ are related by the
Clebsch-Gordan coefficients to the corresponding one-electron
characteristics of the tight-binding approximation. Information
on the relevant values of the parameter ε^Λ can be obtained from
the data on free atoms (for example, from atomic spectroscopy).
The energy spectrum of this system is described by the correlation Hamiltonian and can be calculated quite easily with the aid of
the two-time Green's functions using the method for decoupling
a chain of equations suggested by Hubbard [64] and retaining all
the terms which represent interactions at a single lattice site.
The resultant quasiparticle spectrum is given by the expression

$$E_n(\mathbf{k}) = (\epsilon_{n+1}^{\Lambda_1} - \epsilon_n^{\Lambda_2}) + A\,[\epsilon_0(\mathbf{k}) - \epsilon_d\,], \tag{5}$$

where ε_n^Λ is the energy parameter of the d^n configuration in the
$|\Lambda)$ state; $\varepsilon_0(\mathbf{k})$ is the dispersion law calculated in the standard
one-electron approximation; ε_d is the binding energy of an electron in the d shell; A is the coefficient which can be expressed in
terms of the Clebsch—Gordan coefficients and which depends on
the number of vacancies ("holes") in the d band.

Equation (5) includes explicit information on the electron struc-
ture, which is represented by the dispersion law $\varepsilon_0(\mathbf{k})$ of the standard band approximation. However, the energy spectrum is re-
fined by the inclusion of the configurational stability, which is
represented by the first term in Eq. (5). Consequently, the standard unperturbed d band splits into ten subbands corresponding to
each of the d^n configurations. If the d band is sufficiently wide
and the correlational splitting is weak, the edges of the subbands
become blurred and the usual one-electron band is obtained. In
the opposite case of strong correlation and zero band width, the
excitation spectrum represents a set of levels with energies equal
to the difference between the energies of the states with the d^{n+1}
and d^n configurations. Figure 2 shows the structure of the energy
spectrum in the approximation of fluctuating configurations. We
can see from this figure that the configurational model is inter-

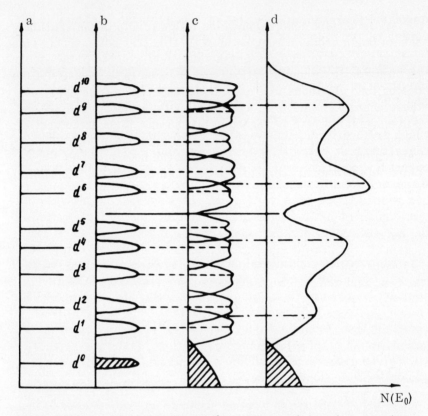

Fig. 2. Structure of the energy spectra of transition metals in various approximations: a) zero-band-width approximation, i.e., energy levels of free atoms as used
in the Heitler—London model; b) narrow-band approximation (model of fluctuating configurations); c) intermediate-band-width approximation (real structures); d)
wide-band approximation conventional band model).

mediate between the Heitler—London and the band models but it
is closer to the Heitler—London case. Since the correlational effects are quite strong in d-type metals (this follows from direct
calculations of the energy parameter of the correlation* and from
the observed variations of the physicochemical properties of the
elements in the long periods), it follows that an allowance for the
many-band configurational nature of the spectrum is often more

---

*The values of the correlation parameter have not yet been calculated for the case
when the screening of the interactions by the delocalized electrons is important [6].

important than detailed information on the actual structure of the energy bands. Equation (5) is dominated by the first term, i.e., the intratomic interactions in solids are of primary importance in the configurational model. Therefore, this model restores the genetic relationship between the free and bound or condensed states of an element (this relationship is lost in the standard band theory). In particular, the exchange correlations to the energy which are obtained on the assumption of a uniform charge distribution (these correlations are used in the standard band theory) cannot be applied to narrow d bands with a strong correlation. However, a qualitative allowance for the interatomic exchange can be made quite simply. This allowance leads to the concept of the stability of electron configurations. The physical origin of the highest energy stability of the half-filled configurations can be described as follows [112]. Electrons in the ground state of the d^5 half-filled configuration have the highest spin moment and the spin component of the total wave function is symmetric. The general condition for the antisymmetry of the wave function of an electron configuration is equivalent to the requirement for the antisymmetry of the coordinate component of the $\psi(r_1, ..., r_n)$ of the wave function. This means that $\psi(r_1, ..., r_n)$ vanishes whenever $r_i = r_j$ (provided $i \neq j$), i.e., the electrons in a spin-symmetric shell tend to "avoid" one another. Consequently, the Coulomb repulsion becomes weaker and the energy of the system decreases. Hence, it follows that the elements containing six or eleven s + d valence electrons are more likely to form the $d^{n+1}s^1$ states than the $d^n s^2$ configurations. Moreover, the additional stabilization is responsible for the maxima of the difference $(\varepsilon_{n+1}^{\Lambda_1} - \varepsilon_n^{\Lambda_2})$ in Eq. (5), which correspond to $n_d = 0$, 5, or 10.

It must be stressed that the highest stability of the d^0, d^5, and d^{10} configurations does not necessarily imply that the one-electron energies of the corresponding d levels are minimal. The values of the one-electron energies do not generally determine the behavior of the d electrons in excitation processes [472]. A quantity of much greater physical significance is the derivative $\partial \varepsilon_{L,S}/\partial n_d$, where $\varepsilon_{L,S}$ is the energy of a configuration in the $|L, S)$ state and n_d is the number of electrons in the d shell. This derivative is calculated on the assumption that the number of electrons in the other shells remains constant. In the condensed state, a large number of multiplets coexist in a single configuration. There-

fore, electron configurations in solids can conveniently be described by the quantity

$$E_d' = \frac{\partial \epsilon_{av}}{\partial n_d}, \tag{6}$$

where ϵ_{av} is the energy of an electron configuration averaged out
over all its multiplets or over the multiplets with the highest possible spin moment. The condition for stability of an atom containing s and d electrons is

$$E_s' = E_d'. \tag{7}$$

A similar condition is applicable also to a system comprising
A and B atoms in the case when electron exchange occurs between
their valence shells. In this case, the definition of Eq. (6) is analogous to Mulliken's definition of the electronegativity.

The charge transferred from the i-th to the j-th shell (i and j
represent the s and d shells of A atoms or the d shells of the A
and B atoms) can be calculated if we know the functional dependence $E_i'(n_j)$. Although this dependence is not yet known for most
of the elements, we can use the equilibrium values of E_i' for different atoms to establish the direction of the charge transfer between the i-th and j-th shells. We can easily show that in the exchange of electrons between the s and d shells of the A atoms or

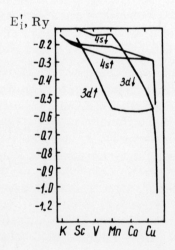

Fig. 3. Variation of the E_i'
energies along rows of the
3d-type element.

between the d shells of the A and B atoms the charge has the tendency to shift to the shells with lower values of E_i', since the condition $E_i' - E_j' < 0$ implies the lowering of the energy.

The recently obtained values of E_i' for the spin-polarized orbitals $s\uparrow$, $s\downarrow$, $d\uparrow$, and $d\downarrow$ in the $d^n s^2$ configurations are shown in Fig. 3. It follows from the curves in Fig. 3 that the energy of a scandium atom, averaged out over its higher multiplets, passes through a minimum when $n_d < 1$, whereas in the case of titanium (and particularly vanadium) some of the s electrons are transferred to the d shell (this is necessary to satisfy the conditions $E_{s\downarrow}' = E_{d\uparrow}'$) and, consequently, $n_s < 2$. In chromium, the spin-up orbital is completely filled and, therefore, its electron configuration is close to $d^5 s^1$. Beginning from manganese, which has the $d^5 s^2$ configuration, and ending with copper (d^{10}), the transfer of some of the s electrons to the d shell becomes increasingly stronger, in the same way as in the first half of this period. Similar data are lacking for the elements in the second and third long periods but we may assume that the basic relationships governing electron transfer are still the same.

The above discussion can be generalized by stating that the transition metals in the condensed state always exhibit a tendency, governed by the energy considerations, to form the most stable d^0, d^5, and d^{10} configurations, which are characterized by the smallest values of E_d'. This tendency results either from a change in the density of the s electrons (particularly in metals) or from the transfer of some of the d electrons from atoms of the second component with a higher value of E_d' (this happens in many-component systems).

It follows from Fig. 3 that the derivative of the energy $\partial E_{av}/\partial n_d$, which represents the energy stability of the various configurations, decreases almost linearly from the d^1 to the d^5 states and from the d^6 to the d^{10} states. In view of this, we can correlate the properties of the elements with the stability of their electron configurations by the use of the statistical weights of atoms with stable configurations (SWASC) of the d^0, d^5, and d^{10} type, calculated on the assumption that the d^1-d^4 and d^6-d^9 intermediate states are absent. These statistical weights provide a measure of how close the average configuration of a metal is to the nearest stable state. The values of the statistical weights of the stable configurations are given by the formulas [91]:

$$P_o = \frac{5-n_d}{5} \left.\begin{array}{c} \\ \end{array}\right\} \quad (8)$$

$$P_s = \frac{n_d}{5} \left.\begin{array}{c} \\ \end{array}\right\} \quad (n_d \leqslant 5), \quad (9)$$

$$P_s = \frac{10-n_d}{5} \left.\begin{array}{c} \\ \end{array}\right\} \quad (10)$$

$$P_{10} = \frac{n_d-5}{5} \left.\begin{array}{c} \\ \end{array}\right\} \quad (n_d > 5). \quad (11)$$

The values of these statistical weights, deduced from the experimental data on the localized valence electrons [43-49, 51, 92], are listed in Table 2.

A similar situation is encountered also in substances formed from sp-type elements. When a solid is formed from free atoms, the outer (highly unstable) shells undergo considerable changes. For example, the s^2p^2 state of a free atom transforms easily, by the $s \to p$ transfer, into the sp^3 configuration in the condensed

TABLE 2. Statistical Weights
(SWASC) of Stable d^0, d^5,
and d^{10} Configurations
in Transition Metals

Atomic number	Element	SWASC, %		
		d^0	d^5	d^{10}
21	Sc	84	16	0
22	Ti	57	43	0
23	V	37	63	0
24	Cr	27	73	0
26	Fe	0	46	54
27	Co	0	18	82
28	Ni	0	2	98
39	Y	78	22	0
40	Zr	48	52	0
41	Nb	24	76	0
42	Mo	16	84	0
71	La	77	23	0
72	Hf	45	55	0
73	Ta	19	81	0
74	W	4	96	0

state [145]. Such s → p transfer occurs more easily in the spectra dominated by the more stable configurations. Atoms with an outer s^2p shell tend to acquire, by the s → p transfer and electron exchange, the quasistable sp^3 configuration. Atoms with s^2p^3 states (nitrogen, phosphorus) undergo s → p transitions in accordance with the scheme s^2p^3 → sp^4 → sp^3 + p, and thus acquire either the sp^3 configuration (with a loosely bound "excess" p electron) or the stable s^2p^6 configuration, which is formed by the acquisition of additional electrons. These features are manifested very clearly in the relationships governing the formation, stability, and physicochemical properties of many systems with the donor — acceptor interaction.

An explicit allowance for the overlap of the shells of neighboring atoms, which transforms electron configurations into energy bands, should result in the broadening of the configurational spectra. The gaps between the neighboring bands in Fig. 2 should disappear when the band widths become of the same order as the interatomic correlations. Under these conditions, a quantitative description must include not only the many-particle effects allowed for in the configurational model but also the actual band structure of a given substance. In this situation, it is useful (and necessary) to calculate the energy spectra of the transition metals with the aid of the existing one-electron methods and to develop additional computation techniques which would improve the internal consistency. In such computations, the structure of the correlated d band is largely governed by the energy spectrum of the unperturbed band obtained in the Hartree—Fock approximation [this follows from Eq. (5)]. Obviously, a consistent inclusion of the kinetic terms in the Hamiltonian of Eq. (4) is essential if we wish to explain many of the properties which originate from the fine structure of the energy bands. However, it must be stressed that the most important characteristic of the band structure, which is the density of states at the Fermi level, can be interpreted unambiguously in terms of the stability of the d^n configurations which are encountered in the transition metals.

The density of states reflects the tendency of the d electrons to become excited and, therefore, it should be governed by the energy gaps between the levels corresponding to the principal d^n configuration and the nearest excited $d^{n \pm 1}$ states. The depen-

dences $N_0(n_d)$ shown in Fig. 2 can be explained bearing in mind
the correlational splitting of the levels which occurs in the limit
of zero band width. The excited levels nearest to the stable configurations are exceptionally high and therefore N_0 should have
a minima for these levels. It is indeed found that the metals
formed from atoms with the stable d^0, d^5, and d^{10} configurations exhibit
a weak tendency for the excitation of electrons and a low density
of states at the Fermi level. Large energy gaps separate also the
ionized levels from the ground states of the d configurations in
the d^2-d^3 and d^7-d^8 range. This is reflected by secondary "dips"
in the dependence $N_0(n_d)$. In the case of the d^1 configuration, the
ionized d^0 level lies very close to the ground state because the
latter is exceptionally stable. It follows that the excitation of
electrons would be quite easy in the metals whose atoms have the
d^1 configuration. The density of states at the Fermi level and the
properties governed by this density would exhibit characteristic
"peaks," which are indeed observed in all such cases. Similar
peaks due to the proximity of the energy of the ground-state configuration to the stable d^5 state are exhibited also by elements
with four or six valence electrons and also by elements with close
to nine electrons (this is typical of the nickel group of metals).
In the case of metals with nine electrons, the ease with which excitation can be achieved is due to the exceptional stability and the
high statistical weight of the d^{10} configurations, which appear
against the background of the d^9 states.

When the temperature is raised, the energy spectrum of a
system should be affected by the entropy factor in the expression
for the free energy $F = E - TS$. However, there is a more important mechanism which results in the redistribution of the probabilities of the various configurations in the spectrum. The $(n+1)s$
shells in the transition metals are adjacent to the nd shells so
that the s electrons participating in various physical processes
frequently compete with the d electrons. When the temperature is
raised, the thermal s → d or d → s excitation effects become increasingly important and the direction of the prevalent excitation
can be deduced from the data on the energy terms corresponding
to the $d^n s^2$, $d^{n+1} s^1$, and d^{n+2} configurations. A special feature of
the d-type elements is that the $d^n s^2$ states are frequently the lowest whereas the $d^{n+1} s^1$ and d^{n+2} states are excited. In the case
of elements located near the middle and the ends of the long periods, the higher energy stability of the d^5 and d^{10} configurations

gives rise not to the $d^n s^2$ but to the $d^{n+1}s^1$ states. In the case of scandium, whose atoms have the $d^1 s^2$ configuration, the s → d transfer is likely at sufficiently high temperatures: such transfer increases the number of electrons in the d shell and reduces the number of s electrons. This applies even more strongly to titanium and vanadium, in which the s shell is completely filled in the free state. In the case of chromium, whose ground state corresponds to the $d^5 s^1$ configuration, we find that the d → s transfer may reduce the number of electrons in the d shell. In manganese, the sequence in which the energy levels are filled reverts to the original scheme and resembles that found in the elements of groups III-VI. At the ends of the long periods, we again encounter a reversal of the positions of the s and d levels which gives rise to a tendency for the d → s excitation and for reduction of the number of the d electrons.

The suggested sequence of filling of the levels in the various configurations in solids is based on the corresponding sequence which applies to free atoms, with the exception of nickel whose configuration in the condensed phase is very close to the stable d^{10} state with a characteristic thermal d → s excitation (of the type encountered in copper). This is not surprising, because the number of d electrons and their spatial distribution in the transition elements should not change greatly in the transition from the free

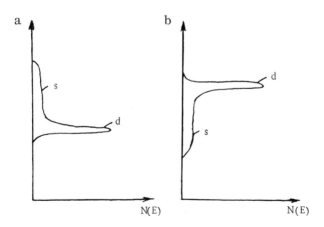

Fig. 4. Schematic distribution of the density of the hybridized sd states: a) transition metals with highest statistical weights of the d^5 and d^{10} configurations; b) other transition metals.

to the condensed state. A more rigorous many-electron approach utilizing the Hamiltonian of Eq. (1) predicts the same features in the energy spectra, except that the concept of the s and d functions should be replaced by the predominant s or d nature of the total functions resulting from the hybridization of the quasilocalized d^n states with the s band. It follows from our discussion that the influence of temperature on the energy spectra of the metals characterized by the highest values of the statistical weights of the d^5 or the d^{10} configurations is quite different from the influence of temperature on the other metals (Fig. 4). The intensity of the s → d or d → s transfer is governed by the hybridization parameter, i.e., the relative value of the third term in Eq. (1), which is of the order of 1-2 eV [473].

The data obtained by direct measurements confirmed these predictions in the case of nickel [455]. The number of the d "holes" in the energy band of this element increases from ~0.6 per atom at low temperatures to a value larger than 1 per atom at T ≈ 1200°K. Recent x-ray spectroscopic data [474] indicate that the number of the d electrons in iron increases when the temperature is raised. Korsunskii, Genkin, and Zavodinskii [475] used the x-ray spectroscopic method in a study of the electron structure of metallic zirconium. An analysis of the profiles and intensities of the x-ray bands of zirconium obtained at low temperatures (α phase) yielded an estimate of 2.6 electrons/atom for the localized fraction of electrons, whereas the corresponding high-temperature measurements carried out on the β phase gave 3.3 ± 0.2 electrons/atom.

We shall show that the opportunity to make an allowance for the influence of temperature is an important advantage of the configurational model over the band description. In the standard band theory, it is difficult to forecast the changes in the electron structure and the properties of metals which should result from the thermal excitation of electrons from the Fermi level. This is the consequence of the indeterminacy of the relative positions and the widths of the s and d bands, and of the relative insensitivity of the Fermi distribution to changes in the temperature.* On the other

---

*The results of recent fundamental calculations of the electron energy spectra of the transition metals at high temperatures and pressures [476] are in agreement with the conclusions drawn from the configurational model.

hand, such predictions can be made quite easily if we use the Heitler — London approximation and this facility is largely retained in the configurational model.

In contrast to the delocalized (collective-state) electrons, the statistical weight of the bonding configurations determines the properties which are controlled by the binding (cohesive) energy. The detailed features of the binding forces acting between atoms in a crystal are not fully reflected in the Hamiltonian of Eq. (4) because the terms corresponding to the interatomic correlation are not included in their correct form. According to the configurational model, the binding forces in substances with identical statistical weights of the d^5 configurations should increase as the stability of these configurations improves, i.e., they should increase with ascending values of the principal quantum number of the d electrons. The d^0 and d^{10} configurations can be regarded as the atomic states. The tendency to form these configurations in a crystal enhances the localization of electrons near the lattice sites (in the case of the d^{10} configurations) or increases the tendency for the delocalization (in the case of the d^0 configurations). In both cases, the interatomic forces are weakened because of the reduction in the number of unpaired d electrons capable of forming bonds.

The nature of the configurational spectra of the transition metals is such that the bonding mechanism is different for metals in the first and second halves of the long periods. If $n_d \leq 5$, the binding forces between atoms result from the overlap of the wave functions of the d^5 configurations. The probability density of these functions has an approximately spherical distribution and the overlap between the nearest neighbors has the nature of a singlet, i.e., the overlap occurs between orbitals with antiparallel spins. This point is stressed by Goodenough: "With relatively small overlap of localized orbitals, the probability of electron transfer is small so that the electrons remain essentially localized; however, the superexchange stabilization introduces antiferromagnetic coupling and an incipient covalency. With relatively large orbital overlap, there is a large probability of electron transfer so that the electrons at any atom are essentially spin-paired" (p. 69 in [60]). This point of view is supported also by other arguments. Jumps of the localized electrons cannot contribute significantly to the metallic bonding because a considerable increase in the energy of the whole system would result from the presence, near a given site, of an electron with spin antiparallel to the spin of the atom located at that site [64]. The complete identity of the atoms in each unit cell and the electrical neutrality of every cell exclude the possibility of ionic bonding. Moreover, the half-filled shells have a natural tendency for the pairing of spins and the formation of covalent bonds. It is stated in [42] that the relationships governing deformation of bcc materials are similar to the relationships observed for covalent crystals. Similar comments are made also in [96].

If $n_d > 5$, the interatomic bonding results in parallel-spin configurations. In some cases, when the gain in stability as a result of resonance compensates the increase in the energy resulting from the change in the spin orientation, the bonding is of the type found in the hydrogen molecule and the d electrons in excess of five per atom are transferred to the collective-state or the localized $s^x p^y$ states. In crystals of the transition metals and, particularly, their compounds, we encounter not only d but also other electron configurations. Therefore, the approach developed here can be generalized to electrons of any symmetry provided sufficient information is available on the existence of the localized component of the electron system. When compounds are formed between the group IV-VI transition metals and the nonmetals of the second period, the d electrons of the metal can be excited to the 2p states to provide stronger $Me-X$ bonds in which the p electrons of the nonmetal component participate. This is supported by theoretical considerations which show that the dsp hybridization is likely in the voids of the cubic and hexagonal lattices [97] and by direct measurements of the x-ray emission and absorption bands of the d-type metals. The $K\beta_5$ band of these compounds is, in fact, due to some p-state admixture and it is this band which is associated with $Me-X$ bonds.

The existence of a definite statistical weight of the $s^x p^y$ configurations also follows from the general theory. Since x-ray spectroscopic investigations and theoretical calculations (the latter carried out by the augmented plane wave method [98]) show that the localized fraction of electrons in compounds is largely of the p-type, it follows that some of the electrons of the p-type symmetry spread into a spectrum in which the $s^x p^y$ configurations predominate.

The increase in the statistical weight of the sp^3 configurations, resulting from the excitation of electrons into the p states adjacent to the 4s states [99] is manifested most clearly by the group IV transition metals in which a considerable fraction of the valence electrons is bound weakly to the atoms. As we go over from titanium to chromium, from zirconium to molybdenum, and from hafnium to tungsten, we find that when the configuration of a free atom approaches the stable form the probability of such excitation becomes less and therefore the Me$-$X bonds in MeX compounds become weaker, whereas the Me$-$Me bonds become stronger. The antibonding orbitals may be filled with the collective-state electrons [57] and this weakens additionally the bonds between the metal and nonmetal components. No comparative calculations have yet been made of the x-ray bonds of all the compounds of the group IV-VI transition metals and, therefore, a direct confirmation of the above hypotheses has not yet been obtained. However, we must stress that these hypotheses make it possible to explain the nature of practically all the published data on the physicochemical, physical, and engineering properties of refractory compounds.

2. Heterodesmic Bonds in Crystals

The heterodesmic nature of bonds in crystals can be included approximately in any of the following models: the atomic model in which all the electrons are assumed to be localized near a given atom or ion; the valence band model in which electrons are assumed to be shared between the nearest neighbors in covalent bonds; and the conduction band model, in which the outer electrons are assumed to be shared by the whole lattice.

In the quantum theory, the binding together of atoms to form a crystal can be regarded as a reduction in the energy of a system of free atoms as a result of their transition to the condensed state. A quantitative criterion of this reduction is

$$\Delta E = -E + \sum_i E_i, \qquad (12)$$

where ΔE is the binding energy of the atoms in a crystal; E_i is the solution of the Schrödinger equation for the particles in question (in this case, atoms); E is the solution of the Schrödinger equation for the complete system with a suitable allowance for all the interactions.

The bonding is called metallic if the greatest reduction in the energy (the greatest increase in ΔE) corresponds to the collective state of the valence electrons. It follows from this definition that the binding energy of alkali metals is basically due to metallic bonding [100, 101].

The formation of chemical bonds as a result of the interaction between atoms alters the states of all the electrons, particularly the valence electrons. In many crystals, including those of the transition metals, the free-energy minimum is attained as the result of considerable localization of the valence electrons. This strong localization in a crystal, corresponding to a high statistical weight of the half-filled states, is a sign of the covalent bonding, which is enhanced by the increase in the energy stability of the electron configurations. From this point of view, the metallic bonding can be regarded as due to collective-state valence electrons, and the ionic bonding as due to the maximum statistical weight of the most stable $s^2 p^6$ configurations of the components of a compound (more rarely due to the combinations of the s^2 and $s^2 p^6$ configurations, as observed in $BeCl_2$).

Depending on the structure of the outer shells of free atoms, which governs the possibility of the formation of stable configurations of the localized valence electrons, all the chemical elements can be divided into three main classes from the electron point of view [36]:

(1) the s elements, which have s outer electrons and completely filled or completely empty deeper electron shells;

(2) the ds and fds elements, whose atoms have partly filled d or fd shells;

(3) the sp elements, which have sp outer electrons.

The first class includes metals of the copper and zinc subgroups, the alkali and alkaline-earth metals, as well as beryllium and magnesium.

The s metals of the copper subgroup (copper, silver, and gold) have the $d^{10} s^1$ configuration in the free atomic state. When metal crystals are formed, the high probability of the appearance of the stable d^{10} configurations and the relatively stable s^2 pairs is responsible for the "noble" properties of these metals and their low electrical resistivities. The increase in the statistical weight and

the stability of the d^{10} states on going over from copper to gold
is accompanied by an increase in the chemical inertness because
the d shell is most susceptible to the capture of electrons. On the
other hand, the d^{10} configurations cannot make a significant contribution to the bonding. The binding forces in these metals are
primarily due to the s electrons. The reduction in the energy stability of the s states on going from copper to gold is accompanied
by a fall in Young's modulus and other characteristics governed
by the low-temperature strength of the bonds.

The metals belonging to the zinc subgroup are formed from
atoms with $d^{10}s^2$ valence shells, which are modified in the condensed
state primarily as a result of the excitation of the s^2 configurations and the resultant formation of a high density of the antibonding delocalized electrons. This is why the melting points of
these metals are lower and their electrical resistivities higher
than the corresponding properties of the copper subgroup.

In the case of alkali metals, the free atoms have the s configurations of the valence electrons. Consequently, the stable configurations are of the s^2 type and their stability (like that of the sp
configurations) decreases with ascending values of the principal
quantum number of the valence electrons. The first ionization
potentials decrease monotonically from lithium to cesium (5.39
eV for Li, 5.138 eV for Na, 4.339 eV for K, 4.176 eV for Rb, and
3.893 eV for Cs) and this is followed by a small rise in the case
of francium (3.98 eV). The ionization potential falls quite strongly
between sodium and potassium, which is evidently due to the partial transfer of the valence electrons to the completely empty 3d
shell of potassium and the elements that follow it in the periodic
table. This is accompanied by changes in the s^2 configurations
and by easier ionization of the atoms. In the case of francium, an
increase in the energy stability of the d states, resulting from the
s—d transitions, raises the ionization energy. A similar effect is
observed for the melting point, which falls from lithium (453.7°C)
to francium (300°C).

The outer electron configuration of free atoms of magnesium
and the alkaline-earth metals is s^2 and these elements share many
properties with nonmetals. This is a consequence of the fact that
the s → p transitions are allowed and the s^2 configurations can be
converted to the sp states, which are typical of nonmetals.

The energy stability of the $s^x p^y$ configurations increases with decreasing values of the principal quantum number of the sp valence electrons. Therefore, the probability of the s → p transitions is particularly high for beryllium and the $s^2 \rightleftharpoons$ sp equilibrium should shift to the right when the temperature is raised because of the somewhat higher energy stability of the sp configuration compared with the stability of the s^2 states. Since the s^2 states govern the metallic properties and the sp configurations are responsible for nonmetallic properties, we find that beryllium exhibits both types of property. In particular, beryllium forms a hydroxide as well as beryllates. The presence of atoms with the sp configurations raises the melting point of beryllium to a value which is one of the highest among metals. The sp configurations in beryllium favor electron capture and the formation of the more stable sp^2, and, particularly, sp^3 configurations. Consequently, beryllium is extremely likely to capture various impurities (which are then difficult to remove) and contaminated beryllium is highly brittle because the electron configurations in the contaminated state are similar to those found in diamond, silicon, or germanium [166].

As we go over to magnesium, the reduction in the energy stability of the sp configurations shifts the $s^2 \rightleftharpoons$ sp equilibrium to the left and this enhances the metallic properties, which are stronger than those of beryllium, although some of the nonmetallic properties still remain (they are due to the presence of atoms with the sp configurations).

The gradual loss of the nonmetallic properties in the case of calcium and other alkaline-earth metals is strongly disturbed by the appearance of the completely empty 3d shell to which the s electrons may be transferred. Therefore, the electron transition equilibrium becomes d $\rightleftharpoons s^2 \rightleftharpoons$ sp. Therefore, the reduction in the statistical weight of the $s^x p^y$ configurations in the alkaline-earth metals is due not only to the reduction in the energy stability of these configurations with increasing principal quantum number but also to the s → d transitions. Since the energy stability of the electron configurations resulting from transitions to the d levels increases with increasing principal quantum number, the equilibrium shifts in the direction of the s → d transitions as we go from calcium to strontium, barium, and radium, and there is a corresponding reduction in the statistical weight of the $s^x p^y$

configurations. The metallic properties of these elements become more pronounced along the same sequence.

These considerations confirm the validity of the separation
of the alkaline-earth metals (with the exception of beryllium and
magnesium), which have many nonmetallic properties, from the
elements of the II-A subgroup in the periodic table.

The second class of elements comprises the sp nonmetals
and semimetals. The most important factors which govern their
physical and chemical properties are the statistical weight and
the stability of the electron configurations.

The increase in the statistical weight and stability of the $s^x p^y$
configurations in the transition from beryllium to carbon is accompanied by an increase in the melting point, formation of special electron configurations, and a reduction in the statistical
weight of the delocalized electrons.

When the principal quantum number of the sp valence electrons increases, the localization of these electrons becomes
weaker, the statistical weight of the stable configurations becomes
less, and the fraction of the delocalized electrons increases. As
we go from carbon (diamond) to lead, the melting point falls, the
microhardness decreases and the forbidden band width diminishes
sharply: 5.2 eV (diamond), 1.21 eV (Si), 0.78 eV (Ge), 0.1 eV (Sn),
and 0 eV (Pb). Nevertheless, lead can be regarded not as a metal
but also as an sp element with a low statistical weight of the sp
configurations, i.e., it can be regarded as a degenerate semiconductor. The decrease in the value of statistical weight of the sp
configurations results in a change in the structure of lead compared with the structures of diamond, silicon, germanium and
gray tin. In the case of gray tin, the statistical weight of the sp^3
configurations, which are necessary for the formation of the diamond-like lattice, becomes sufficiently high at temperatures below −17°C. At higher temperatures, particularly at room temperature, this material transforms into white tin with a lower
statistical weight of the sp configurations and a higher density of
the delocalized electrons. It is possible that a sufficiently high
density of the sp configurations is formed also in lead at very low
temperatures and this would give rise to a diamond-like structure
and semiconducting properties.

In the case of boron, the statistical weight of the sp^3 configurations is significantly lower than the corresponding weight for diamond and this is why the melting point, forbidden band width, and microhardness have lower values than in the case of diamond. As we go over to aluminum, the stability of the sp configurations decreases even more strongly and the statistical weight of the delocalized electrons increases. This is also observed for gallium, indium, and thallium, which, for all practical purposes, are metals. The electron configurations of these elements are stabilized when they form compounds with nitrogen and its analogs: this happens because of the acquisition of an additional electron.

When diamond-like $A^{III}B^V$ compounds are formed (here, A = B, Al, Ga, In, Tl; B = N, P, As, Sb, Bi), an atom of the element B with an electron configuration of the s^2p^3 type undergoes a transition ($s^2p^3 \rightarrow sp^4 \rightarrow sp^3 + p$) which gives rise to sp^3 and p electrons and the latter are transferred to the element A. Consequently, the s^2p configuration is transformed to the sp^2 state when a compound is formed and the latter state is converted to the sp^3 configuration by the acquisition of one p electron. Thus, the sp^3 configurations are formed in the crystal lattices of these compounds and the stability of these configurations is highest when the principal quantum numbers of the s and p electrons are lowest. Therefore, the compound BN has the widest forbidden band and the highest melting point in the diamond-like borazon modification. As the principal quantum number of the components A and B increases, the stability of the sp^3 configurations falls, the exchange of electrons between the components is enhanced, and, consequently, the values of the melting point, microhardness, and forbidden band width all decrease (Table 3).

In the case of nitrogen and its analogs, we encounter both the sp^3 and s^2p^6 configurations. In particular, the nitrogen atom can give up one p electron to form a carbon-like configuration

$$s^2p \rightarrow sp^4 \rightarrow sp^3 + p.$$

This transition is responsible for the formation of diatomic nitrogen molecules which have the $sp^{3+x} : sp^{3+x}$ configuration and are strongly bound.

The capture of electrons and the completion of the p shell are more pronounced in oxygen than in nitrogen and, therefore, oxygen has a higher chemical activity. However, the transfer of elec-

TABLE 3. Physical Properties of $A^{III}B^{V}$ Semiconductors

Compound	Melting point, °C	Forbidden band width, eV	Microhardness, kg/mm²
BN (borazon)	3000	5	10 (Mohs' scale)
AlN	2200	3.8	9 (" ")
GaN	1500	3.25	—
InN	1200	—	—
BN	2500	5.9	3700
AlP	1800	2.92	—
GaP	1467	2.25	940
InP	1058	1.34	410—430
BAs	2000	3	1900
AlAs	1700	2.16	500

trons and transitions of the type encountered in nitrogen can still occur in oxygen.

The statistical weight of the stable s^2p^6 configurations of the localized valence electrons increases in the elements of group VII. The transition from chlorine to bromine reduces the stability of the s^2p^6 configurations and gives rise to the liquid state under normal conditions. An even stronger deterioration in the stability of the s^2p^6 configurations is observed for iodine, which is solid at the normal temperature and pressure (the forbidden band width of this element is 1.3 eV).

The inert gases, which have stable and practically noninteracting s^2p^6 configurations, are normally monatomic. The energy stability of these configurations decreases with increasing values of the principal quantum number of the sp "valence" electrons but it is quite high. Therefore, of the inert gases, only krypton, xenon, and radom form compounds and these compounds are usually fluorides (fluorine exhibits the strongest tendency for the capture of an electron and the formation of the stable s^2p^6 configuration) [102-104]. The compounds of xenon with oxygen are also known [104]. Hence, we can postulate the existence of krypton and radon oxides and of radon and xenon chlorides.

The transition metals with partly filled d shells (in the free state) occupy a special position. Several models have been put forward to explain the electron structure of these elements. A satisfactory description of these metals and of refractory compounds is provided by the configurational model. This is best illustrated by considering the physicochemical properties of these substances.

CHAPTER II

Physicochemical Properties of Transition Metals and of Their Alloys and Compounds

3. Crystal Structure and Polymorphism

The crystal lattice symmetry is one of the most important characteristics of a solid and in many ways is related to the physicochemical properties of the solid. The symmetry in the distribution of atoms in a crystal is responsible for the crystallinity of a solid and distinguishes it from, for example, amorphous materials and liquids. A change in the lattice symmetry is accompanied by changes in such properties as the density, specific heat, conductivity, magnetism, hardness, strength, and plasticity. Therefore, the relationships governing the formation of the crystal structures of the transition metals are important from the theoretical and practical points of view.

In general, the stability of a crystal structure is governed by its free energy. Under any given set of conditions the most stable is that modification which has the lowest free energy. The difficulties encountered in the calculation of the free energy at T = 0°K (this energy is equal to the internal energy except for the contribution of the zero-point vibrations) make theoretical computations impractical. The difference between the internal energies of different crystalline modifications is slight and its order of magnitude is close to the errors associated with the calculations of thermodynamic properties. At T > 0°K we have the additional problem of calculation of the traces of the operators which occur in the expression for the partition function.

33

It is simpler to construct physical models which take account of the special features of the electron structure of the transition metals and of its changes-with temperature. The first attempts of this kind were made by Hume-Rothery [29] and Jones [30]. Later developments included a more rigorous allowance for the localization of electrons in the d-type metals. Many relationships are explained by Pauling [68] and Grigorovich [113], who used variants of the theory of valence bonds. The next steps were made by Goodenough [171], Korsunskii and Genkin [172], who made allowance for the presence of the localized and collective-state d electrons. Engel [77] and Brewer [82-83] recently concluded that it would be necessary to make an allowance for the special features of the d^n configurations obtained in each case and for the higher stability of the d^5 state. Engel postulated different degrees of the s → d transfer in pure transition metals and assumed that polymorphic transitions were due to the transfer of atoms from one quantum state to another, as a result of an increase in the temperature. In this way, Engel attempted to extend to the transition metals the Hume-Rothery rules which were inapplicable in their original form because of the polyvalence of the d metals.

Brewer's diagrams are empirical and can be used in studies of complex systems. However, they do not prove the validity of Engel's postulates. For example, the Engel−Brewer rule is not obeyed by some of the elements with relatively simple electron structures [173]. The success of the Engel−Brewer model is due to an allowance for the partial localization of electrons and for the stability of the d^5 configurations. The weaknesses of this model follow from the assumption that unpaired electrons participate in the bonding process but do not affect the nature of the crystal structure which is assumed to be governed solely by the sp electrons. Since many of those properties which are governed by the sp electrons in simple (nontransition) metals are determined by the localized d electrons in the transition metals, it would be surprising if the d electrons had no influence on the lattice symmetry.

In the configurational model the stability of the d^0, d^5, and d^{10} states affects the nature of the structures corresponding to the beginning, middle, and end of each of the long periods [174]. In the hypothetical case of 100% content of the d^0 or d^{10} spherically symmetric stable configurations, we would have lattices with the highest symmetry. This circumstance, together with the tendency to achieve the highest packing density, should give rise to fcc structures. This type of lattice is typical also of completely filled or completely empty configurations of other types.

The half-filled stable configurations are associated with structures whose existence can also be predicted on the basis of the configurational model because the nature of these structures is related to the maximum value of the spin in these half-filled configurations. The greatest reduction in the energy of a system can be expected for the configurations with antiparallel spins separated by shorter distances than those which occur in the parallel-spin states. Two sublattices of atoms with different directions of spins can occur in bcc structures because the spins of neighboring atoms are always antiparallel. This cannot occur in fcc structures because their characteristic feature is the zero-spin configuration (in such structures the two nearest neighbors of a given atom can be nearest neighbors to each other, i.e., these structures have higher energies than bcc lattices with nonzero-spin configurations). Moreover, the lower statistical weights of the d^5 and d^{10} configurations and the higher statistical weights of the intermediate configurations favor the formation of lower symmetry structures. The high statistical weight of the intermediate configurations as well as the higher fraction of the collective-state electrons tend to favor the formation of hcp structures. Similar ideas have been put forward by Cornish [175] and Slater [176].

We can now see why the low-temperature sequence of the lattice symmetry starts with fcc structures for the alkaline-earth metals at the beginning of a period, passes through hcp and bcc structures for groups V and VI, and goes back to hcp and fcc for elements at the ends of the long periods. The c/a ratio for hcp structures is usually less than the theoretical value of 1.633 (Table 4).

Structures of this type are typical of the group IV transition metals (titanium, zirconium, and hafnium), which are characterized by fairly low statistical weights of the d^0 (57, 48, 45%) and d^5 (43, 52, 55%) configurations. These metals do not have high-symmetry lattices and the high density of the collective-state

TABLE 4. Crystal Structure of Some Elements (c/a ratio of hcp structures is given in parentheses)

IA (s^1)	IIA (s^2)	IIIA (d^1s^2)	IVA (d^2s^2)	VA (d^3s^2, d^4s^1)	VIA (d^5s, d^4s^2)	VIIA (d^5s^2, d^6s^1)	VIIIA (d^6s^2, d^7s^1)	VIIIB (d^7s^2, d^8s^1)	VIIIC (d^8s^2, d^{10}, d^9s^1)	IB ($d^{10}s^1$)
K bcc — 0—337	Ca fcc — 0—737 hcp (1.63) 737—1123	Sc hcp (1.585—8) 0—1223 bcc (1223—1811)	Ti hcp 1.586 0—1158 bcc 1158—1938	V bcc — 0—2190	Cr bcc — 0—2176	Mn A12 — 0—1000 A 13 1000—1365 tetr-fcc 1365—1406 bcc 1406—1517	Fe bcc — 0—1183 fcc 1183—1663 bcc 1663—1912	Co hcp (1.623) 0—700 fcc 700—1763	Ni fcc — 0—1728	Cu fcc — 0—1356
Rb bcc — 0.312	Sr fcc — 0—506 hcp (1.63) 500—813 bcc 813—1043	Y hcp (1.585—8) 0—1763 bcc — 1763—1773	Zr hcp (1.502) 0—1135 bcc — 1135—2128	Nb bcc — 0—2770	Mo bcc — 0—2890	Tc hcp (1.604) 0—2473	hcp (1.583) 0—2700	Rh fcc — 0—2239	Pd fcc — 0—1823	Ag fcc — 0—1234
Cs bcc — 0—302	Ba bcc — 0—983	La hcp (1.61) 0—583 fcc 583—1137 bcc 1137—1193	Hf hcp (1.587) 0—2050 bcc —	Ta bcc — 0—3270	W bcc — 0—3650	Re hcp (1.615) 0—3308	Os hcp (1.579) 0—3500	Ir fcc — 0—2727	Pt fcc — 0—2043	Au fcc — 0—1136

sp electrons favors the formation of hcp structures. The non-zero spin moment of the intermediate configurations stabilizes hcp structures with c/a < 1.633, so that the nearest neighbors can have the antiparallel spin orientation and the second-nearest the parallel orientation. These structures are exhibited also by scandium and yttrium which are characterized by high values of the statistical weight of the intermediate d configurations and high values of the density of the delocalized sp electrons. The transition to metals in groups V and VI results in an increase of the statistical weight of the d^5 configurations in vanadium, chromium and its analogs. Therefore, these elements crystallize in bcc structures. Since the statistical weight of the d^5 configurations in hafnium is 55% and that in vanadium 63%, we may assume that the bcc structure becomes stable when the statistical weight of the d^5 states reaches about 60%.

This has recently been confirmed by the results of an x-ray spectroscopic investigation of the valence electron distribution in β-Zr at T $\approx$ 1500°K, carried out by M. I. Korsunskii and Ya. E. Genkin. It was established that the number of the localized bonding electrons in β-Zr is 3.3 ± 0.2 and the number of the collective-state electrons is 0.7 ± 0.2, whereas in α-Zr these numbers are 2.6 ± 0.2 and 1.4 ± 0.2, respectively, i.e., the localization increases with increasing temperature and the statistical weight of the d^5 configurations in β-Zr is of the order of 66%.

No data are available on the localized valence electrons in manganese. This is due to the very complex electron structure of this element. The anomalous properties of manganese can be attributed to the high density of the sp electrons (which are formed as a result of dissociation of the s^2 pairs, typical of free manganese atoms) and to the high statistical weight of the intermediate configurations which "loosen" the lattice and reduce its symmetry.

Iron is characterized by a low statistical weight of the intermediate configurations which, together with the low content of the collective-state sp electrons, cannot stabilize an hcp structure in spite of the fact that the statistical weights of the d^5 and d^{10} states are quite low (54 and 46%, respectively). This distribution of electrons in iron is responsible for the low-temperature bcc structure. Since the statistical weight of the d^5 state in cobalt is lower than that in iron, the lattice becomes hcp and this is primarily due to the participation of the intermediate configurations and of the collective-state sp electrons. In the case of copper and nickel the fcc structure is primarily due to the d^{10} configurations. The other metals of the copper subgroup (silver and gold) also have the fcc structure due to the participation of the d^{10} configurations which are pushed apart slightly by the sp electrons.

The hcp structure with $c/a < 1.633$ is the "compromise" symmetry between the bcc and fcc structures which are formed in metals with high statistical weights of the d^5 and d^{10} configurations.

The reduction in the free energy is due to enhancement of the metallic bonding which results from reduction in the atomic volume. The deviation of c/a from its ideal value permits a somewhat larger separation between atoms with parallel spins than between atoms with antiparallel spins and this results in an additional gain in the binding energy as a result of covalent bonding in such crystals.

The dissociation of the s^2 pairs, typical of the zinc subgroup, gives rise to a high density of the collective-state electrons in crystals of these metals. Therefore, the lattices of these metals have low symmetries: they are hexagonal in the case of zinc and cadmium and rhombohedral in the case of mercury. The highest statistical weight of the d^0 configuration is observed for the alkaline-earth metals and this enhances the stability of the fcc structures.

In contrast to iron, which is characterized by the bcc structure at low temperatures (this structure is associated with the d^5 states), ruthenium and osmium crystallize in the hcp modification although the $4d^5$ and $5d^5$ configurations are more stable than the $3d^5$ state. As we go over from molybdenum to ruthenium and from tungsten to osmium, we encounter an additional electron in the stable d^5 configuration of free atoms. This electron increases strongly the energy of free atoms (this can be seen from the ionization potential curves in Fig. 1). Consequently, when such atoms form solids, the excess electrons are transferred to the sp levels and this increases the statistical weight of the d^5 configuration, and enhances the collective state in the crystal. The high statistical weight of the d^5 configuration in a crystal combined with a considerable fraction of the delocalized electrons tends to favor hcp structures, which are observed for technetium, rhenium, ruthenium, and osmium. These are followed by fcc modifications in the case of VIIIB and VIIIC subgroups. These modifications are due to an increase in the statistical weight of the d^{10} configurations and a fairly high density of the delocalized electrons in these elements.

When a transition metal is heated, it often undergoes polymorphic transitions which are determined by the nature of the changes in the configuration spectrum with temperature. When the temperature is raised, the statistical weight of a given configuration and the d localization in a crystal both increase, with the exception of metals in the VIA and VIIIC subgroups, for which the d → s excitation reduces the statistical weight of the stable d^5 or d^{10} states. This change in the statistical weight is reponsible for the appearance of polymorphic transitions and it explains the basic tendency observed in polymorphism: the low-temperature α-type modification of most of the polymorphic metals (Li, Na, Ca, Sr, Ba, Sc, Y, La, Ac, and Ti, Zr, Hf, and others — a total of 30 metals) has an hcp or fcc structure which changes to a bcc form when the temperature is raised because the statistical weight of the d^5 configuration reaches its maximum value. Polymorphism is not exhibited by some of the elements characterized by the highest values of the statistical weights of atoms with the stable configurations (this group includes the inert gases, copper, silver, gold, zinc, radium, and mercury). No modifications have been found in the group VIII metals with sufficiently high statistical weights of the d^{10} configuration: they include palladium, iridium, ruthenium, and osmium. The exceptions to this rule are iron, cobalt, and nickel, which have lower statistical weights of the stable configurations than their analogs in this group. This happens because the stability of the $3d^5$ configurations is less than that of the $4d^5$ and $5d^5$ states. On the other hand, in the case of group IV elements (titanium, zirconium and hafnium), which are characterized by a low statistical weight of the d^5 configuration and hexagonal symmetry at low temperatures, the increase of this statistical weight with increasing temperature may give rise to the hcp → bcc transition. The existence of this transition is evidence that the statistical weight of the d^5 configurations of these metals increases (with increasing temperature) to about 60%.

The polymorphic transition temperatures of titanium, zirconium, and hafnium are 1156, 1035, and 2033°K (all these values are taken from [177]). This sequence is due to the following circumstances. In the case of zirconium, the stability of the d^5 states is higher than in titanium because of the larger quantum number of the valence electrons, which favors stabilization of the β structure. Further increase of the stability of the d^5 configurations in hafnium should result in an additional fall of the polymorphic transition temperature. However, the filled 4f shell and the associated increase of the energy gap between the s and d levels hinder the s $\rightarrow$ d excitation at higher temperatures and reduces the rate of growth of the statistical weight of the d^5 configuration. Therefore, the "critical" statistical weight of the d^5 configurations in hafnium is reached at much higher temperatures than in zirconium.

The statistical weight of the stable d^5 configurations in the transition metals of groups V and VI (vanadium, niobium, tantalum, chromium, molybdenum, and tungsten) is in excess of 60% at room temperature. Further increase of the temperature can only reduce it but not below 60%. Thus, all these should have (and actually have) only one crystalline modification, which is associated with the high statistical weight of the stable configurations and which has the bcc symmetry. This is in agreement with the experimental results of V. N. Svechnikov et al., who demonstrated the absence of polymorphic transitions in pure chromium and rejected the earlier postulate of numerous modifications of the metal. It is also doubtful whether the β modification of pure tungsten does exist and it is likely that the fcc structure of this metal mentioned in [178] is the result of stabilization by a few impurities (such stabilization can occur at moderate temperatures).

Manganese is characterized by a large number of polymorphic transitions: This is due to the existence of localized atomic magnetic moments (resulting from the low stability of the d^5 states), the high statistical weight of the intermediate configurations, and the presence of the delocalized sp electrons which interact with the localized moments. The presence of the d^5 states in manganese gives rise to simple cubic α and β modifications and to a tetragonal face-centered γ modification. Technetium and rhenium are characterized by a higher stability of the d^5 configurations and by a Mott transition in the electron subsystem. These two metals have hcp structures.

Iron is characterized by even lower values of the statistical weight of the d^5 configuration and of the intermediate configurations, by a small fraction of the sp electrons, and by the existence of localized magnetic moments which do not favor the appearance of an hcp structure. Consequently, iron exists in two modifications which are almost equally stable (fcc and bcc structures) and the competition between these structures is responsible for the nature of the polymorphic transitions in iron.* The reduction in the importance of the d^5 configurations in the formation of the lattice gives rise to an hcp modification in the case of cobalt, which is followed by a transition to an fcc structure because the statistical weight of the d^{10} configuration increases with temperature. In the case of nickel and copper, the d^5 configurations are even less important and, therefore, these metals have the fcc structure throughout the range of existence of the solid phase. The temperatures of the polymorphic transitions decrease along the sequence iron — cobalt — nickel because of the increasing ease of the formation of the d^{10} configurations and the associated tendency to form fcc structures.

The position is similar in the case of platinoids, except for the fact that the higher proportion of the collective-state electrons in these metals and the increasing stability of the d^5 configurations favor the Mott transition, the disappearance of localized moments (this hinders polymorphic transitions), and the stabilization of the hcp structure, which is observed also in rhodium and which changes to the fcc form at higher temperatures because of the increase in the statistical weight of the d^{10} configuration. Palladium has the fcc structure at all temperatures right up to the melting point because the stability of the d^{10} configurations in this metal is much higher than in nickel. In the third long period, the aforementioned relationships governing polymorphic transitions and their temperatures are similar to those found in metals in the second long period.

Scandium and yttrium belong to the transition-metal group and are characterized by a high statistical weight of the inter-

---

*The high-temperature $\gamma \rightarrow \delta$ transition in iron is due to an increase in the entropy term in the free energy, whose value is highest for the bcc structure.

mediate configurations responsible for the existence of hcp structures in these metals at low temperatures. When the temperature is raised, the statistical weight of the intermediate states decreases and the corresponding weight of the d^5 increases. This stimulates transitions of these elements to bcc modifications. Their analog, lanthanum, has three modifications, of which the α-modification (with the hcp lattice) is associated with the lower statistical weight of the d^5 configuration and the considerable "loosening" of the lattice by the collective-state electrons and by the intermediate configuration spectrum.

Further increase in the temperature should raise the statistical weight of the d^5 configurations and favor a transition to the bcc modification. However, the high-lying unfilled f shell plays an important role in lanthanum. Thermal excitation at some temperatures can even reduce the statistical weight of the d^5 configurations of lanthanum. At 310°K the fcc modification becomes more stable because of an increase in the statistical weight of the d^0 configuration. A significant rise of the statistical weight of the d^5 configurations at higher temperatures results from the s → d excitation and it favors the formation of the bcc structure based on these configurations.

Scandium, yttrium, and lanthanum are often grouped with lanthanoids as rare-earth metals although, in fact, they are typical d-type elements and members of the rows of these elements in the first, second, and third long periods. The f-type elements (Table 5), whose free atoms have f-shell electrons, exhibit a tendency for the f → d transitions [179, 180], which is governed by a curve representing all possible terms: the probability of the f → d transitions decreases when the number of possible terms increases. Since the d^5 configurations are energetically more stable than the f^7 configurations, it follows that the structure should be governed by the statistical weight of the d^5 configurations resulting from the f → d transitions. It is evident from Fig. 5 that the temperatures of the polymorphic transitions in the high-temperature modification region increase from cerium to samarium, i.e., they increase with decreasing probability of the f → d transitions and with increasing difficulty of attainment of sufficiently high statistical weights of the d^5 configurations on which the bcc structure is based.

TABLE 5. Polymorphic Transitions in Rare-Earth Elements

Element	Modification	Crystal structure	Transition temperature, °C	Valence electron configuration
Ce	α	fcc	−103 $(\alpha \to \gamma)$	$4f^25d^06s^2$
	β	hcp	90 $(\beta \to \gamma)$	
	γ	fcc	730 $(\gamma \to \delta)$	
	δ	bcc	—	
Pr	α	hcp	792 $(\alpha \to \beta)$	$4f^35d^06s^2$
	β	bcc		
Nd	α	hcp	862 $(\alpha \to \beta)$	$4f^45d^06s^2$
	β	bcc		
Sm	α	rhombo-hedral	917 $(\alpha \to \beta)$	$4f^65d^06s^2$
	β	bcc		
Eu	α	»	—	$4f^75d^06s^2$
Gd	α	hcp	1264 $(\alpha \to \beta)$	$4f^75d^16s^2$
	β	bcc		
Tb	α	hcp	1317 $(\alpha \to \beta)$	$4f^95d^06s^2$
	β	cubic		
Dy	α	hcp	—	$4f^{10}5d^06s^2$
Ho	α	»		$4f^{11}5d^06s^2$
Er	α	»	—	$4f^{12}5d^06s^2$
Tm	α	»	—	$4f^{13}5d^06s^2$
Yb	α	fcc	798 $(\alpha \to \beta)$	$4f^{14}5d^06s^2$
	β	bcc		
Lu	α	hcp	—	$4f^{14}5d^16s^2$

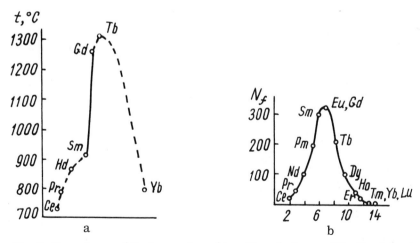

Fig. 5. Dependences of the temperature of transition of a bcc modification (a) and of the number of possible terms N_f (b) on the number of electrons in the f shells of free atoms of the rare-earth metals.

Polymorphic transitions do not occur in europium, whose free atoms have the f^7 configuration and which do not undergo the f → d transitions (the f^7 configuration gives rise, like the d^5 state, to bcc structures). The statistical weight of the stable states in this element is so high at ordinary temperatures that any further increase in the temperature causes no polymorphic transitions.

In the case of gadolinium, whose atoms have a 5d electron, we again have a definite probability of the f → d transitions resulting in the formation of the d^5 states and a departure from the f^7 configuration. The high stability of the d^5 configurations is responsible for the high temperature of the polymorphic transition from the lower to the higher symmetry. This is also observed for terbium, whose configuration is not $4d^9 5d^0 6s^2$ but $4d^8 5d^1 6s^2$, as demonstrated in [181] on the basis of a study of thermionic emission from terbium compounds.

No polymorphic transitions are observed between dysprosium and thulium (inclusive) and these metals are known only in the hcp modification which forms because of the high density of the collective-state electrons and the moderately high statistical weight of the d^5 configuration. This monomorphic series breaks at ytterbium, whose low-temperature fcc structure, associated with the high value of the statistical weight of the f^{14} configuration, transforms to the bcc structure at higher temperatures. Finally, lutetium exhibits only the hexagonal structure associated with the presence of a d electron additional to the f^{14} configuration, as in the case of gadolinium and terbium. In view of this, it is very likely that there is a high-temperature bcc modification of lutetium.

It is worth noting that the $\alpha \to \beta$ transition temperature of ytterbium is much lower than the corresponding transition temperatures of gadolinium and terbium in view of the high probability of the f → d transitions. Clearly, high-temperature modifications should also be observed for Dy, Er, Ho, and Tm. The temperatures of the $\alpha \to \beta$ transitions in these metals should decrease in a regular manner from dysprosium to thulium provided the temperatures sufficient for the excitation of the f^7 and f^{14} configurations and for their transformation into the d^5 configuration lie below the melting points of these elements.

An important conclusion which follows from the interpretation of polymorphism suggested in [174] is that polymorphic transitions are due to changes in the statistical weights of the stable and intermediate configurations and changes in the density of the delocalized electrons in a crystal, irrespective of the cause of these changes. This is in agreement with the modern views on polymorphism which is regarded just as any other structural change in which the chemical composition is preserved. This conclusion also allows us to determine the nature of polymorphism of the transition metals under high pressures. An allowance should be made for the tendency to form stable electron configurations, and for the conditions of the formation and changes in these configurations (in particular, changes due to partial transitions of the electrons to deeper levels).

Numerous experimental data [183] show clearly that the behavior of matter at high pressures, particularly such an important characteristic as the compressibility, is governed by the outer electrons in the atomic shells [184]. The probability of electron transitions, considered from the point of view of the formation of stable electron configurations, is not the same within a given group in the periodic system and this applies even more strongly for neighboring groups.

In the case of the transition metals, one ought to note the changes in the compressibility of those metals whose free atoms contain fewer than five electrons in the d shell.

For example, titanium is characterized by a low statistical weight of the d^5 configurations, a weak binding of atoms in the lattice, and a strong tendency for the s → d transitions. Under pressure, these transitions increase the statistical weight of the bonding configurations and the binding forces, which tends to reduce the atomic spacing in titanium. In the case of vanadium and, particularly, chromium, the statistical weight of the d^5 configuration is high even under normal conditions and there is limited scope for increasing it. Therefore, the compressibility decreases rapidly along the Ti → V → Cr series. This applies also to the Zr → Nb → Mo and Hf → Ta → W series for which the compressibility generally decreases with increasing principal quantum number of the d electrons, which is to be expected because of the increasing stability of the d^5 configurations.

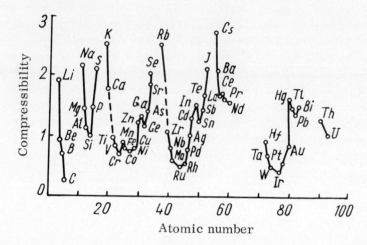

Fig. 6. Compressibility of the elements.

The compressibility of manganese is quite high (Fig. 6) because of the re-formation of the pairs of the s^2 and d^5 configurations under pressure (these pairs are partly dissociated in the formation of crystals from free atoms). Therefore, we can expect rhenium also to have a considerable compressibility but, in this case, the anomaly should be much less because of the much higher stability of the d^5 states in rhenium compared with manganese (under normal pressure).

A very high compressibility is expected for scandium which is characterized by a low value of the statistical weight of the d^5 configuration and a corresponding large scope for increasing this weight under pressure. A characteristic feature of all the elements with low values of the statistical weight of the d^5 state is the relatively large contribution of the s and s^2 configurations to the bonding (this contribution is comparable with that made by the d electrons). Consequently, the compressibility and other properties vary in a characteristic manner in the cesium, titanium, copper subgroups, and partly in the nickel subgroup. Since the stability of the s and s^2 configurations decreases with increasing principal quantum number of the s valence electrons, the transitions from cesium to yttrium, from titanium to zirconium, and from copper to silver are all accompanied by an increase in the compressibility as a result of higher strength of the bonding between the lattice atoms. Further transition to the group III elements of the third long period (La, Hf, and Au) results in a large fall of the stability of the s configurations and in a decrease of their contribution to the bonding. However, the stability of the d^n configurations increases and in these elements the latter dominates the strength of the bonding. Consequently, the compressibility of La, Hf, Pt, and Au is lower than the compressibility of their analogs in the second long period.[*]

In the case of elements belonging to the groups ranging from V to VIIIB, the nonmonotonic variation of the compressibility disappears and is replaced by a regular dependence when elements with higher principal quantum numbers of the valence electrons are considered. This is due to an increase in the statistical weight of the d^5 configurations and the greater importance of these configurations in the general bonding scheme.

Since the statistical weights of the d^0 configurations and the fraction of the delocalized electrons are high for the group IV elements (Ti, Zr, Hf) under normal conditions and since the d^5 configurations begin to predominate under suitable excitation conditions, we may assume that the temperatures of the $\alpha \rightarrow \beta$ polymorphic transitions of these metals should decrease with increasing pressure because of the rise in the statistical weights of the d^5 states. At sufficiently high pressures, the hexagonal α modification should disappear completely or it should exist in a very narrow range of temperatures.

In the transition metals with more than five electrons in the d shell of free atoms the statistical weight of the d^{10} configurations increases with increasing number of electrons in the d shell. Since the d^{10} state is the limiting configuration which these elements can reach under compression, it follows that an increase in the statistical weight of the d^{10} configurations under normal conditions should be accompanied by a reduction in the compressibility. On the other hand, a reduction in the statistical weight of the d^5 states reduces the binding forces, increases the atomic spacings, and, consequently, provides a margin for an

---

[*]In the case of the group VIIIC elements these two factors are compensated in the second period: the compressibility of nickel is equal to the compressibility of palladium but higher than the compressibility of platinum.

increase in the compressibility at the expense of changes in the crystal structure. Thus, in the first half-period we can expect a regular fall in the compressibility because the total statistical weight of the stable configurations and the statistical weight of the bonding configurations both increase from group IVA to group VIA. In the second half-period, the variation of the compressibility should be more complex because an increase in the total statistical weight of the stable d^5 and d^{10} configurations and a simultaneous fall in the weight of the bonding configurations are observed between the metals of group VIIIA and the metals of group VIIIC. This is favored also by the higher stability of the d^{10} configurations as compared with the d^0 states. The compressibility of ruthenium and rhodium decreases somewhat compared with molybdenum and that of osmium and iridium decreases in comparison with tungsten (Fig. 6). Similar relationships are observed for the properties which represent the strength of atomic bonding at low temperatures and are a consequence of the high stability of the $4d^5$ and $5d^5$ configurations. This stability is responsible for the "expulsion" of the excess (i.e., left over after the formation of the d^5 configurations) electrons to the sp levels. This raises the statistical weight of the d^5 states in the system to a value close to that observed for the group VI metals and gives rise to an additional increase in the binding energy compared with molybdenum and tungsten because of the finite statistical weight of the $s^x p^y$ configurations formed by these "expelled" electrons. This effect is hardly observed in the first long period because of the low stability of the 3d configurations.

The strength of the bonds and the compressibility increase strongly along the sequences Co → Ni → Cu, Rh → Pd → Ag, Ir → Pt → Au because of the increase in the atomic spacing as a result of the reduction in the statistical weight of the bonding configurations. This can be established quite easily by comparing Figs. 6 and 7, which show changes in the atomic spacing for elements in the long periods. Moreover, a general fall in the compressibility is observed in any given group as a result of an increase in the stability of the d^5 configurations because of an increase in the principal quantum number of the d valence shell.

Lanthanoids and actinoids exhibit f → d transitions when metal crystals are formed from free atoms. The probability of these transitions decreases with increasing number of the pos-

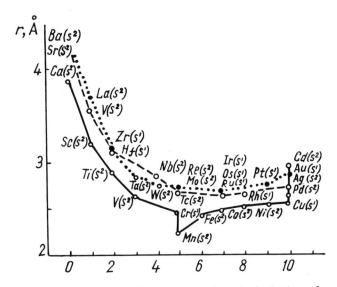

Fig. 7. Dependence of the atomic spacing r in the lattices of d-type transition metals on the number of electrons in the d shell of free atoms.

sible terms (Fig. 5), i.e., this probability is highest for cerium (lanthanum has no f electrons) and it decreases toward gadolinium. The application of pressure should give rise to transitions of the s and d electrons to the f levels. The probability of such transitions should also decrease with increasing number of possible terms. In other words, as in the case of the d-type transition metals, the compressibility should increase when the state of the f electrons shifts away from the stable f^7 and f^{14} configurations (the f^0 state does not form). It follows from the experimental results (Fig. 6) that the compressibility does indeed decrease from cerium to praseodymium and then to neodymium as the statistical weight of the f^7 configurations in metal crystals increases under normal conditions. We may postulate that the compressibility will decrease further right down to gadolinium and then will vary in accordance with a law approximately similar to that obeyed by platinoids, i.e., the law will be affected by the existence of the collective-state electrons. It is reported in [185] that an electron transition is indeed observed in cerium at relatively moderate pressures (1.243×10^4 kg/cm^2).

It follows from this interpretation that we can expect irreversible formation of new compounds at high pressures because of the stabilization of the outer shells of atoms. For example, it is very probable that compounds can form between niobium ($4d^4 5s^1$) and technetium ($4d^6 5s^1$), between yttrium ($4d^1 5s^2$) and rhodium ($4d^8 5s^1$), and many other elements as a result of high-pressure-induced increase in the statistical weight of the stable configurations, which would be retained after the removal of pressure.

The problem of the influence of alloying on the nature and temperature of polymorphic transitions in the elements is also of importance. The considerable theoretical interest evinced recently in the problem of alloying (in relation to the general problem of the influence of impurities on the electron structure of solids) and the important practical applications (search for elements suitable for the formation of refractory alloys) make it necessary to consider this problem with the aid of the configurational model. This is particularly important because such criteria as the volume factor, electronegativity, isostructural nature of the components, and electron density do not provide a satisfactory basis for the physical theory of alloying.

It follows from the general aspects of the configurational model that an impurity atom introduced into a metal crystal must change more than a matrix atom in order to reestablish the original bonding with the surrounding atoms. These changes involve modification of the electron configuration of the impurity in such a way that it becomes similar to the average configuration of atoms in the matrix.* This is the basic difference between alloying and the formation of alloys and compounds because, in the latter case, the concentration of atoms of the second component is sufficiently high to result in the formation of a second configurational spectrum. Thus, a study of the influence of admixtures in titanium on the temperature of the polymorphic $\alpha \rightarrow \beta$ transition has established that all the d metals and copper reduce the temperature of this transition at a rate $\partial T / \partial c$ (c is the impurity concentration) which increases along the Nb−Mo and Ni−Cu sequences, i.e., it increases with increasing difference between the electron structures of the host and impurity atoms (Table 6). We may as-

---

*This conclusion is supported by the relationships governing changes in the magnetic moments of impurities in ferromagnetic transition metals [58].

TABLE 6. Influence of
Alloying on the $\alpha \rightarrow \beta$
Transition Temperature
in Titanium

Metal	Conc. of alloying element, at.%	Transition temperature, °C
Ti	—	882
Ni	0.2	850
	0.4	840
	1.0	800
Cu	0.2	870
	0.4	860
	1.0	830
	1.5	800
Nb	0.5	855
	0.9	830
	1.5	810
Mo	0.3	860
Al	4.0	940

sume that the forced transfer of electrons from the impurity to the titanium atoms increases the statistical weight of the d^5 configurations responsible for the formation of the bcc structure and favors the appearance of this structure at lower temperatures. Similar conclusions have been reached by many other workers [187].

The more complex relationships obtained for the $\alpha \rightarrow \gamma$ transition in iron are associated with the special features of the electron structure of this element [188].

A free atom of iron has the outer valence electron configuration $3d^6 4s^2$. In the solid state below 910°C iron exists in the bcc α-Fe modification; in the 910-1100°C range we have the fcc γ-Fe phase and above 1400°C the bcc δ-Fe structure is observed.

Depending on the nature of their neighbors, the atoms of iron in a solid either loss some of their valence electrons which are then used mainly to form the stable d^5 configurations or they acquire electrons from their neighbors and tend to form the stable d^{10} configurations (the first process is more likely).

The influence of the transition metals on the nature of solid solutions of iron can be seen from Table 7 [9]. It is evident from

TABLE 7. Influence of Various Elements on Solid Solutions of Iron

Element	Valence electron configuration in free atoms	Crystal structure		α-Iron	γ-Iron
		at room temperature	at high temperatures		
Ca	$3d^{0}4s^{2}$	fcc	hcp, bcc	Insoluble	Insoluble
Sc	$3d^{1}4s^{2}$	fcc, hcp	bcc	Extends very narrow range of α solid solutions	—
Ti	$3d^{2}4s^{2}$	hcp	»	Extends fairly wide range of α solid solutions	—
V	$3d^{3}4s^{2}$	bcc	»	The same	—
Cr	$3d^{5}4s^{1}$	»	»	Extends range of α solid solutions at high temperatures	Extends narrow range of γ solid solutions (up to 8 at.%)
Mn	$3d^{5}4s^{2}$	Complex bcc	fcc, complex bcc	—	Forms continuous series of solid solutions and extends range of γ solutions
Fe	$3d^{6}4s^{2}$	bcc	fcc, bcc	—	—
Co	$3d^{7}4s^{2}$	hcp	fcc	Extends fairly wide range of α solid solutions (up to 50 at.%)	Extends fairly wide range of γ solid solutions (above 50 at.%)
Ni	$3d^{8}4s^{2}$	fcc	»	—	Forms continuous series of solid solutions and extends range of γ solutions
Cu	$3d^{10}4s^{1}$	»	»	Extends narrow range of α solid solutions	Extends narrow range of γ solid solutions
Zn	$3d^{10}4s^{2}$	hcp	hcp	Extends range of α solid solutions (up to 42 at.%)	Reduces range of γ solid solutions
Sr	$4d^{0}5s^{2}$	fcc	hcp, bcc	Insoluble	Insoluble
Y	$4d^{1}5s^{2}$	hcp	bcc		

Zr	$4d^2 5s^2$	↗	↗	Extends fairly wide range of α solid solutions	—
Nb	$4d^4 5s^1$	bcc	↗	The same	—
Mo	$4d^5 5s^1$	↗	↗	↗	—
Tc	$4d^5 5s^2$	hcp	hcp	↗	Extends fairly wide range of γ solid solutions
Ru	$d^7 s^1$	hcp	hcp	↗	Extends fairly wide range of γ solid solutions (up to 20 at.%)
Rh	$d^8 s^1$	fcc	fcc	Extends range of δ solid solutions at high temperatures	Forms continuous series of solid solutions and extends range of γ solutions
Pd	$d^{10} s^0$	↗	fcc		Practically insoluble
Ag	$d^{10} s^1$	↗	hcp	Practically insoluble	The same
Cd	$d^{10} s^3$	hcp	bcc	The same	Insoluble
Ba	$5d^0 6s^2$	bcc	fcc, bcc	Insoluble	↗
La	$d^1 s^2$	hcp	bcc	↗	—
Hf	$d^2 s^2$	↗		Extends narrow range of α solid solutions	—
Ta	$d^3 s^3$	bcc	↗	Extends fairly wide range of α solid solutions	↗
W	$d^4 s^2$		↗	The same	Extends range of γ solid solutions (up to 10 at.%)
Re	$d^5 s^2$	hcp	hcp	Extends range of α solid solutions (at high temperatures extends δ range)	
Os	$d^6 s^2$	↗	↗	The same	Extends fairly wide range of γ solid solutions
Ir	$d^7 s^2$	fcc	fcc		The same
Pt	$d^9 s^1$	↗	↗		Extends narrow range of γ solid solutions
Au	$d^{10} s^1$	↗	↗	Extends narrow range of α solid solutions	Practically insoluble
Hg	$d^{10} s^3$	rhombohedral		Practically insoluble	

Notes. There have been reports of the existence of a low-temperature modification of vanadium, of polymorphic transitions in ruthenium at 1035, 1190, and 1500°C, and of a transition in rhenium at 1000–1200°C. The atomic radii of calcium, strontium, barium, silver, and mercury do not satisfy the Hume-Rothery rule.

this table that Ca, Sr, Ba, Y, La, Ag, and Cd as well as Hg do not dissolve in iron. This is due to the fact that the dimensions of the atoms and ions of these elements are fairly large and, moreover, they have the d^0 or d^{10} and s^2 configurations in the localized fraction of the valence electrons. The probability of formation of the highly stable d^5 states in combination with the iron atoms is very low for these impurities. The dimensions of the atoms and ions of the listed elements do not satisfy the Hume-Rothery rule, which says that only those elements can dissolve in iron whose atoms differ by not more than $\pm 15\%$ from the dimensions of the iron atom.

Although the dimensions of the scandium, zirconium and hafnium atoms also lie outside the range of atomic radii satisfying the Hume-Rothery rule, the probability of formation of the d^5 configurations for these elements is higher than for Ca, Sr, Ba, Y, and La and they form solid solutions with iron.

It is also clear from Table 7 that Sc, Ti, V, Cr, Zr, Nb, Mo, Tc, Hf, Ta, W, and Re extend the range of the α-type solid solutions. The atoms of these elements have either a high statistical weight of the d^5 states in the solid phase (Cr, Mo, Tc, W, Re) or they are characterized by a high probability of the formation of these states from their own electrons (because of the s → d transitions) and from the iron electrons. The atoms of these transition metals have (d + s) electrons fewer than the iron atoms and when they are dissolved in iron the localized fraction of the valence electrons of Fe decreases and the statistical weight of the d^5 states increases, favoring the formation of bcc structures and extension of the range of the α-type solid solutions.

The transition metals which have (d + s) more electrons than iron (Co, Ni, Cu, Ru, Rh, Pd, Os, Ir, Pt, Ag) usually extend the range of the γ-type solid solutions with fcc structures. This is due to the fact that the atoms of these metals transfer some of their (d + s) electrons to the iron atoms so that the statistical weight of the d^5 configurations of iron increases and the statistical weight of the d^{10} states increases.

We have already mentioned that high statistical weights of the d^5 states correspond to bcc structures and fairly high weights of the d^{10} configurations correspond to fcc structures. The influence of the transition metals on the nature of iron-base solid solutions confirms our point of view on the relationship between the stable electron configurations and the nature of the crystal structure.

The transition metals whose atoms have electron configurations similar to those of iron (Cr, Mn, Co, Rh, Re) exert an influence on the nature of iron-base solid solutions which is intermediate between the influence exhibited by the groups mentioned in the preceding paragraphs. This influence is fairly strong in the case of chromium, manganese, and cobalt, which are characterized by an energy stability of the d^5 configurations lower than those exhibited by the transition metals in the other periods. In the free state the atoms of chromium have the valence electron configuration $3d^54s$ and they have $(d + s)$ electrons less than the iron atoms. Chromium extends considerably the fairly wide range of the α-type solid solutions of iron and if present in concentrations up to 8 at.% chromium it also extends the usually narrow range of the γ-type solid solutions. Manganese $(3d^54s^2)$ also has $(d + s)$ electrons less than iron but it extends only the range of the γ-type solid solutions. The behavior of cobalt, which has $(d + s)$ electrons more than iron, is quite interesting: up to 50 at.% it extends the range of the α-type solid solutions and above 50 at.% it extends the range of the γ-type solid solutions. This influence of chromium, manganese, and cobalt is due to the relatively low energy stability of the d^5 configurations of these elements.

The ability to extend the range of the α solid solutions decreases with increasing energy stability of the d configurations and with increasing tendency for "expulsion" of the excess electrons to the sp levels. This applies to the analogs of cobalt, such as rhodium and iridium, which are characterized by nine $(d + s)$ electrons: rhodium can extend the range of δ-type solid solutions with bcc structures at high temperatures, whereas iridium extends only the range of the γ solid solutions with fcc structures.

Ruthenium and osmium have the same number of the valence $(d + s)$ electrons as iron but the stability of the electron configurations in these metals is higher because of the higher value of the principal quantum number of the valence electrons. Therefore, ruthenium and osmium can only extend the range of the γ solid solutions.

Thus, the atoms of the transition metals with less than eight (d + s) electrons, which increase the statistical weight of the d^5 configurations in iron, usually extend the range of the α solid solutions with bcc structures. On the other hand, atoms of the transition metals with eight or more (d + s) electrons, which have a tendency to transfer valence electrons to iron and to increase the statistical weight of the d^{10} configurations of iron, extend the range of the γ solid solutions with fcc structures.

This approach is based on an analysis of the energy stability of the various electron configurations and can be applied to other characteristics associated with the crystallochemical nature of the d and f elements. The electron structure of free atoms determines the stability of the various phases in more complex systems and, in particular, it determines the type of lattice which is observed in various compounds. We shall not consider this point in detail but the reader is referred to [189] for hydrides, [190] for beryllides, [191] for borides, [192] for carbides, [193] for nitrides, [194] for chalcogenides, [195] for magnides, [196] for aluminides, [197] for silicides, and [198] for germanides. These papers should be studied in conjunction with later investigations [199-202], which indicate that in compounds with nonmetals the transition elements act as donors whose ability to give up electrons decreases with increasing statistical weight of the stable configurations and increasing energy stability of the d configurations. Some of the aspects of the crystal chemistry of compounds will be discussed in later sections.

4.   Thermodynamic Properties

A convincing proof of the high stability of the d^5 and sp^3 configurations of atoms in the transition metals, their alloys, and compounds is provided by the observed regularities in the thermodynamic properties of these substances [203].

The atomic binding is most closely related to the heat of sublimation, which represents the energy necessary for the detachment of an atom from a solid and its transition to the gaseous state. The stronger the binding forces in the solid, the higher is the numerical value of the heat of sublimation. The transition from Ca to Sc, Ti, and V in the first long period is associated with an increase in the statistical weight of the d^5 configurations, which results in a tendency for the heat of sublimation to increase

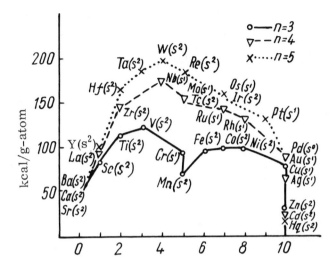

Fig. 8. Heats of sublimation of the d-type transition metals plotted as a function of the number of electrons in the d shell of a free atom.

(Fig. 8). In the case of the group VI metals the statistical weight of the d^5 configurations decreases with increasing temperature and at temperatures close to those needed for sublimation the statistical weight of these configurations may be much lower than the corresponding weight in the group V metals. This is why the heat of sublimation falls suddenly in the case of chromium and manganese (Fig. 8), this tendency in the latter element being enhanced by a pronounced bond-weakening effect of the delocalized electrons (this effect is manifested by the dissociation of the s^2 configuration pairs, which are typical of free manganese atoms).

The statistical weight of the d^5 configurations in iron, cobalt, and nickel is less than that in vanadium and this is why the heats of sublimation of this triad are lower; however, these heats are still somewhat higher than the value for chromium. When the temperature is raised the statistical weight of the d^5 configurations of iron decreases because of the d localization and because of an increase in the statistical weight of the d^{10} configurations. In the case of nickel the effect of temperature is opposite: it reduces the statistical weight of the d^{10} configurations and raises the weight of the d^5 configurations. As a result of such leveling of the statistical weights the heats of sublimation of iron, cobalt and nickel are approximately equal (the value for cobalt is slightly higher because of the "optimal" combination of the aforementioned factors). The heat of sublimation of copper is lower because the statistical weight of the d^{10} states in this element predominates so that the binding energy of copper is governed primarily by the statistical weight of the half-filled s configurations. The heat of sublimation of zinc falls even further for the same reasons as in the case of manganese compared with chromium: the presence of the s^2 configurations, in addition to the stable groups, gives rise to delocalized antibonding electrons in the condensed phase and weakens the atomic binding.

A similar variation in the heat of sublimation is observed between yttrium and palladium. A less marked fall of the heat of sublimation is observed on transition from niobium to molybdenum and technetium; this fall is due to the same factors as in the case of chromium and manganese, except that these factors are weakened by the stabilization of the d^5 configurations with increasing principal quantum number of the valence electrons. An analogous variation in the heat of sublimation is observed between lanthanum and platinum. The highest stability of the d^5 configurations in the case of tungsten is responsible for the statistical weight inequality $d^5(Ta) < d^5(W)$ at all temperatures and it "fills" the dip of the kind observed for chromium and manganese and for molybdenum and technetium.

A consequence of the rise of the stability of the d^5 configurations with increasing principal quantum number of the valence electrons is an increase in the heats of sublimation in certain groups of elements. This increase is observed for all the metal analogs and it is particularly strong when we approach the group VI elements (Cr, Mo, W), for which the statistical weight of the bonding configurations is the highest. Conversely, the stability of the sp configurations falls with increasing principal quantum number of the electrons participating in these configurations. Therefore, we may expect a decrease in the heats of sublimation in the alkaline-earth and noble metal groups and in the zinc group, which is characterized by an almost completely filled d shell.

The heat of sublimation does indeed decrease along the Zn → Cd → Hg sequence and the lowest value of the heat is observed for mercury. This applies also to the heats of sublimation of the elements located between Ca and Sr and between Cu and Ag. A further fall in the stability of the s^2 configurations (in the case of barium) and of the s configurations (in the case of gold) is observed in the third long period. The stability of the d^5 configurations of the elements in this period increases strongly and this enhances the statistical weight of these configurations (particularly at high temperatures). The fall in the stability of the s configurations is masked by the much stronger rise of the stability and the statistical weight of the d^5 states so that the binding energy of the metals in the third long period becomes higher. These metals also exhibit clearly temperature dependences of the thermodynamic characteristics of the type described above (this applies to the characteristics governed by the strength of the binding in the lattice).

If we consider the heats of fusion Q_f of the transition metals (Fig. 9) as characteristics of the properties of these metals at lower temperatures (the melting points of the 3d metals are 1500-2200°K and their boiling points are 3000-3500°K), we find that the dependences plotted in Figs. 8 and 9 are similar: the heats of fusion of these elements increase right up to the middle of a long period and fall toward the end of the period because of similar changes in the statistical weights of the d^5 configurations in all the periods. A similar general variation is observed also within

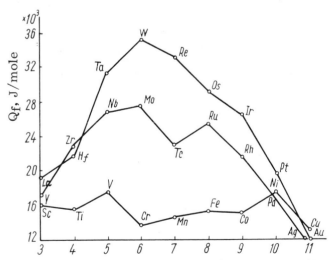

Fig. 9. Heats of fusion of the d-type transition metals plotted as a function of the number of the (s + d) valence electrons.

groups (the value of Q_f increases with increasing principal quantum number of the d valence electrons in similar configurations). However, this analogy between variations of the heats of sublimation and the heats of fusion is valid only to the extent to which we can ignore the temperature dependences of the statistical weights of the d^5 configurations. The following effects are observed when the temperature is lowered sufficiently:

1. The statistical weight of the d^5 configurations decreases considerably in the group V metals and increases in the chromium group. A comparison of Figs. 8 and 9 shows that the difference between the heats of fusion of vanadium and chromium is less than the corresponding difference between the heats of sublimation. In the fifth period the higher stability of the $4d^5$ configurations is responsible for the statistical weight of the d^5 configurations being higher for molybdenum than for niobium. Therefore, $Q_f(Mo) > Q_f(Nb)$, in spite of the fact that the inequality is reversed for the heats of sublimation.

2. The increase in the statistical weight of the d^5 configurations in nickel at high temperatures has a somewhat stronger effect that in the case of cobalt.

3. The heats of sublimation are affected also by the s → d (in the case of Ba) and by the d → s (in the case of Au) excitations. Consequently, the heat of sublimation of gold becomes higher than the corresponding heats of silver and copper, whereas the heat of fusion of gold falls below that of copper but becomes higher than that of silver.

TABLE 8. Debye Temperatures of d-Type Transition Metals

Element	Lattice	Θ, °K	Element	Lattice	Θ, °K	Element	Lattice	Θ, °K
Ti	bcc	300	Zr	bcc	212	Hf	bcc	200
V	»	335	Nb	»	250	Ta	»	230
Cr	»	490	Mo	»	380	W	»	310
Mn	Complex cubic	380	Tc	hcp	—	Re	hcp	275
Fe	bcc	432	Ru	»	400	Os	»	250
Co	fcc	385	Rh	»	370	Ir	»	285
Ni	»	390	Pd	»	290	Pt	»	233

TABLE 9. RMS Thermal Displacements of Atoms $(U_{291°K})$
in d-Type Transition Metals [177]

Element	U_{291}, Å	Element	U_{291}, Å	Element	U_{291}, Å
Ti	0.149	Zr	0.136	Hf	0.127
V	0.123	Nb	0.114	Ta	0.104
Cr	0.113	Mo	0.082	W	0.073
Fe	0.107	Ru	—	Os	—
Co	0.111	Rh	0.078	Ir	0.065
Ni	0.113	Pd	0.134	Pt	0.112
Cu	0.136	Ag	0.156	Au	0.159

Further changes in the configuration spectra can be observed
by considering such characteristics of the d-type metals (which
can be regarded as thermodynamic properties) as the Debye temperature Θ and the rms thermal displacements of atoms U.

When the temperature is lowered from the melting point to the
Debye temperature (200-600°K), the statistical weight of the d^5
configurations in the group VI elements tend to become higher
than the corresponding weights for the group V elements, including the 3d metals. Therefore $\Theta_{Cr} > \Theta_V$ (Table 8). On the
other hand, the statistical weight of the d^5 configurations in nickel
is even higher than that in cobalt and this is why the Debye temperature of nickel is higher. This effect is not manifested by palladium and platinum because of the higher stability of the 4d configurations in these metals. We should also mention that the increasing stability of the 4d configurations in the palladium group[*]
gives rise to a different effect which can be described as follows:
at low temperatures and in the case of sufficiently high stabilities
of the d^5 configurations (this applies to elements in the second and
third long periods) it is found that in metals with six and seven d
valence electrons the expulsion of an excess electron to a p level
is favored by the energy considerations and this increases the
weight of the $s^x p^y$ configurations and gives rise to high values
of the weight of the d^5 states, which approach 100%, as in the case
of molybdenum and tungsten. Therefore, the low-temperature

---

[*] A similar effect should also be manifested in the third long period. The values reported in [157] are far too low, as demonstrated by the opposite effect recorded in
[117], i.e., by the observation that the Debye temperature increases from tungsten
to rhenium and then begins to fall in a regular manner.

binding strength of the group VIIIA and VIIIB metals may be affected by the d^n and $s^x p^y$ configurations so that the binding strength of ruthenium, rhodium, osmium, and iridium can be higher than the corresponding strength in molybdenum and tungsten. At higher temperatures the increasing d localization tends to raise the statistical weight of the d^{10} configurations and reduce the weight of the d^5 states.

Therefore, the high-temperature characteristics of the binding strength (including thermodynamic parameters) have their highest values for molybdenum and tungsten whereas the corresponding low-temperature characteristics are highest for ruthenium and osmium. This effect is observed also for the 4d and 5d metals at Debye temperatures which obey $\Theta(Ru) > \Theta(Rh) > \Theta(Mo)$ and $\Theta(Os) > \Theta(Ir) > \Theta(W)$. At room temperature an effect of this kind can be illustrated by considering the rms displacements of atoms resulting from thermal vibrations. In this case we observe easily the low-temperature effects (Table 9) manifested by $U(Cr) < U(V)$ (i.e., the statistical weight of the d^5 configurations in chromium is higher than in vanadium); $U(Co) < U(Ni)$ (the statistical weight of the d^5 configurations in cobalt is higher than in nickel). Moreover, $U(Rh) < U(Mo)$ and $U(Ir) < U(W)$ (no data are given in [177] for ruthenium and osmium). Finally, the binding strength of the copper group metals at room temperature decreases (U increases) along the series Cu $\rightarrow$ Ag $\rightarrow$ Au, which is evidence of the predominant contribution of the s configurations to the bonding and of the very high value of the statistical weight of the d^5 configurations, i.e., the d $\rightarrow$ s excitation is weak in these metals. It is probable that the binding strength of gold falls below that of silver at higher (probably even Debye) temperatures. However, the fall of the Debye temperature observed in the copper group does not reflect reduction in the binding strength but it is more likely due to an increase in the mass of the atoms. This follows from the observation that the Debye temperature decreases on transition from the 3d to the 5d metals and in group VI. It also follows from the Lindemann approximation for the Debye temperature

$$\Theta \approx 0.137 T_{mp}^{1/2} \rho^{1/3} / M^{5/6}. \tag{13}$$

For these reasons the entropy S of individual substances cannot be used as a measure of the binding strength of the transition metals in a group (its increase is not due to weakening of the binding forces but due to an increase in the mass of the vibrating atoms). On the other hand the atomic masses exert hardly any influence on the entropy in the d-metal series. Therefore, the entropy should decrease with increasing values of the statistical weights of the d^5 configurations, because of increasing order in the crystals (which strengthens the binding forces), which is indeed observed (Fig. 10). The principal contributions to the entropy are made by the "lattice" and "electronic" terms. The nature of changes in the lattice entropy S_L can be analyzed quite simply using the thermodynamic relationship

$$S_L = -\left(\frac{\partial F_L}{\partial T}\right)_V \qquad (14)$$

and the expression for the free energy of a vibrating system of atoms considered in the Debye approximation [75]

$$F = E_0 + N\nu T\left[3\ln(1 - e^{\frac{\Theta}{T}}) - D\left(\frac{\Theta}{T}\right)\right]. \qquad (15)$$

Here, E is the energy of the interaction between atoms at T = 0°K (per mole); ν is the number of atoms per unit cell; Θ is the char-

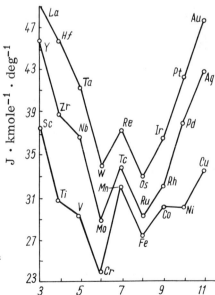

Fig. 10. Entropies of the d-type transition metals at 293°K plotted as a function of the number of the (s + d) valence electrons.

acteristic (Debye) temperature, which is related to the maximum
Debye frequency by the expression $k\Theta = \hbar\omega m$;

$$D\left(\frac{\Theta}{T}\right) = \frac{3}{\left(\frac{\Theta}{T}\right)} \int_0^{\frac{\Theta}{T}} \frac{y^3 dy}{e^y - 1}. \tag{16}$$

In the limiting case of high temperatures $(T > \Theta/4)$ the asymptotic
form of the function F is given by the expression

$$F = E_0 - 3N\nu \ln kT + 3N\nu \ln \hbar\bar{\omega}, \tag{17}$$

where $\bar{\omega}$ is the geometric mean frequency and

$$\ln \bar{\omega} = \frac{1}{3N\nu} \int_0^{\omega_m} \ln \omega d\omega.$$

Differentiation of Eq. (17) gives the following expression for the
lattice entropy:

$$S_L = 3N\nu \ln kT - 3N\nu \ln \frac{\pi\omega}{e} = 3N\nu \left(\ln kT - \frac{\pi\omega}{e}\right). \tag{18}$$

We can easily see that the lattice entropy differs from one
transition metal to another mainly because of differences in $\bar{\omega}$.
In most cases we may assume that $\bar{\omega} \simeq \omega_m$ and, therefore, the
variation of the entropy along a given series of elements should
be "anticorrelated" with the corresponding variation of the Debye
temperature. It is also logical to expect a relationship between
the entropy and Young's modulus or the velocity of sound in a
crystal.

It is also worth considering the electronic component of the
entropy [75]

$$S_e = -\frac{d\Phi}{dT}, \tag{19}$$

where

$$\Phi = \Phi_0 - \frac{\pi^2}{6} V k^2 T^2 N(E_0) \tag{20}$$

is the thermodynamic Gibbs potential; Φ_0 is the contribution of the
electronic part of this potential at $T = 0°K$; $N(E_0)$ is the density of

states at the Fermi level E_0. Differentiation of the Gibbs potential yields

$$S_e = \frac{\pi^2}{3} VN\,(E_0)\,k^2 T. \tag{21}$$

The electronic component of the entropy can be considered in
the two limiting cases of $kT \gg E_0$ and $kT \ll E_0$. Usually the electron gas in a metal can be regarded as degenerate at all temperatures and, therefore, it is the second condition that is normally
satisfied in metals. Numerically, the electronic component of the
entropy is equal to the electronic component of the specific heat
C_e, which is governed directly by the electron structure of a crystal.

The anomalously high values of C_e for the d metals, compared with the nontransition elements, are due to the special position of these metals in the periodic system [204]. The value of γ
in $C_e = \gamma T$ of the d metals is found to oscillate between groups
[205, 206] and with increasing principal quantum number [207].
It is not possible to explain the behavior of the electronic specific
heat by attributing it to the complex nature of the energy bands because this simply replaces one problem with another. A satisfactory solution to this problem may be obtained by allowing for
the influence of the localization of the d valence electrons in the
atomic or bonding states [153]. This reveals the cause of oscilla-
tions of the function $N(E)$, which determines many properties of
the d transition metal.

The variation of the electronic specific heat of metals between
Sc and Cu, between Y and Ag, and between La and Au can be explained with the aid of the configurational model. It is natural to
assume that the localized electrons make no contribution to the
electronic specific heat and that the value of C_e is due to the
presence of the sp and d electrons in the collective state (the principal contribution is made by the d electrons).

These ideas can be combined with the dependence of γ on the
number of the d electrons (Fig. 11) to yield the following information.

I. The deepest minima in the curves for each of the three
long periods are observed for those elements whose free atoms
are characterized by the most stable configurations d^0, d^5, and

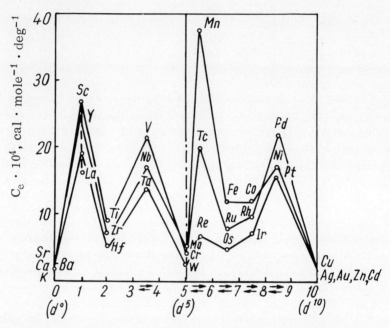

Fig. 11. Electronic specific heats of the d-type transition metals plotted
as a function of the number of electrons in the d shell.

d^{10}. These configurations correspond to the strongest localization of the d electrons and such localized states make the smallest
contribution to the electronic specific heat.

II. The metals which are the nearest neighbors of those described in the preceding paragraph exhibit a sharp rise (a "peak")
of the electronic specific heat. In view of the above considerations, such peaks can be interpreted as follows. A relatively
small change in the number of the d electrons in a metal compared with any of the stable configurations d^0, d^5, or d^{10} results
in a redistribution which tends to produce the largest possible
number of atoms with the electron configurations modified to give
the nearest stable state. The remaining d electrons are transferred to the collective state and they make a large contribution
to the electronic specific heat. We can now distinguish two possible cases: (A) the number of the d electrons in a metal is less
than that necessary for the formation of a stable electron configuration; and (B) the number of electrons is higher than that
needed for the same stable configuration.

Obviously, in the random process of formation of the stable electron configurations in metals and alloys the relative number of the d electrons in the localized and collective states will vary rapidly with the total number of the d electrons per atom in the d metal (case A). This is confirmed by the experimental results (Fig. 11) which show that the slope of the $V-Cr$, $Nb-Mo$, $Ta-W$, $Ni-Cu$, $Pd-Ag$, and $Pt-Au$ lines (measured with respect to the ordinates) corresponding to the d^0, d^5, and d^{10} stable electron configurations is higher than the slope of the $Sc-Ca$, $Y-Sr$, $La-Ba$, $Mn-Cr$, $Tc-Mo$, and $Re-W$ lines [these slopes are, respectively, $(0.2-0.8) \times 10^{-4}$ and $(1.0-1.9) \times 10^{-4}$ cal $\cdot$ mole^{-1} $\cdot$ deg^{-2} $\cdot$ electron^{-1}].

III. A minimum of the electronic specific heat is observed in the middle of each of the intervals d^0-d^5 and d^5-d^{10}, corresponding to the two possible combinations of the stable electron configurations. These minima are weaker than those corresponding exactly to the d^0, d^5, and d^{10} configurations.

We have mentioned that the transfer of the d electrons to the collective state increases when the number of d electrons (n_d) is ~1.4 or ~6.9 and that this effect decreases in the central parts of the intervals d^0-d^5 and d^5-d^{10} because the energy gaps between the levels corresponding to the d^n and d^{n+1} configurations in the first case are smaller than in the second. When n_d is ~1.4 or ~6.9, the statistical weight of the excited configurations increases. This results in a sharp rise of the electron density in manganese, which exhibits a strong tendency for the formation of the collective states of electrons as a result of dissociation of the s^2 pairs which are found in free atoms. Therefore, the anomalously large number of the electrons which can contribute to the specific heat makes its value especially large compared with the specific heats of the other elements in the same period.

The electronic specific heat and similar properties are sensitive to an increase in the statistical weight of all possible stable configurations. Therefore, the periodic dependence of the electronic specific heat on the number of the d electrons is quite different from the corresponding dependence of, for example, the heat of sublimation of the transition metals (this heat is a measure of the strength of the atomic binding and it is sensitive to the population of the bonding d states). We have mentioned earlier that the localized bonding states of the d electrons correspond to the stable d^5 configurations and, therefore, the highest values of the heat of sublimation are observed for those transition metals which are located in the central parts of the transition periods where statistical weight of the d^5 configurations has its highest value. The high values of the statistical weights of the d^0 and d^{10} states in the metals corresponding to the beginning and the end of each of the transition periods does not enhance the atomic binding because these d^0 and d^{10} stable configurations are not the bonding but the localized atomic states. It is worth noting that when the atomic number of the transition metals in a given group increases, i.e., when the principal quantum number of the d shell rises, the electronic specific heat decreases. According to our model, this is due to an increase in the stability of similar configurations with increasing value of the principal quantum number of the d shell, which enhances the effect of the d-electron localization in the transition metals and reduces the number of the collective-state electrons which contribute to the electronic specific heats.

The experimental data on the electronic specific heat of the rare-earth metals can be interpreted in the same way as in the preceding paragraphs. The deep 4f shell of the rare-earths is gradually filled along the series and the distribution of those electrons which are in excess of the xenon configuration can be represented by $4f^m5s^25p^66s^2$, where m increases with increasing atomic number from 0 for lanthanum to 14 for ytterbium and the filling of the 5d shell begins from lutetium. In the condensed phase the rare-earth atoms are usually in the form of triply charged positive ions. The f $\rightarrow$ d transition ($4f^m6s^2 \rightarrow 4f^{m-1} \cdot 5d^16s^2$) gives rise to about three electrons in the conduction band, which is a combination of the overlapping 6s and 5d bands. The probability of the f $\rightarrow$ d transition is particularly high for those rare-earth metals which have the stable f electron configurations $4f^0$, $4f^7$, and $4f^{14}$, i.e., this probability is highest for La, Gd, and Lu. The first and the last of these three metals can be regarded (with equal justification) as a d or an fd transition metal. An analysis of the experimental results [211, 212, 216] given in [216] has established that the electronic specific heat is the same for most of the rare-earth metals (to within $\pm 1\%$) and its value is 25×10^{-4} cal $\cdot$ mole^{-1} $\cdot$ deg^{-2}, i.e., it is of the same order of magnitude as the corresponding specific heat of the d transition metals in group III. This is due to the same dominant factor, which is the presence of a collective-state d electron (the low stability of the d^1 configuration in the presence of a neighboring stable configuration d^0 cannot give rise to localized d electron states). Europium and ytterbium, which correspond to the stable electron configurations f^7 and f^{14}, behave as divalent elements in the condensed phase [216] because of the very low probability of the f → d transitions in these metals (the configuration of the d electrons in europium and ytterbium is close to d^0). Consequently, the electronic specific heat of europium and ytterbium is considerably lower than the corresponding heat of the other rare-earth elements and its value is 13.9×10^{-4} and 6.9×10^{-4} cal · mole^{-1} · deg^{-1}, respectively [217]. Unfortunately, it is not possible to carry out a detailed study of the dependence of the electronic specific heat on the population of the 4f shell because some of the rare earths are ferromagnetic (Pr, Ho, Gd, Nd, etc.) with a strong exchange interaction between the f electron spins. This effect makes a large contribution to the specific heat right down to the lowest temperatures and the electronic component cannot be isolated from the measured values of the specific heat [216, 218, 219].

An examination of the nature of formation of compounds between the transition metals and the nonmetals in the second period shows that chemical bonding in these compounds is complex (heterodesmic) and the Me−Me, Me−X, and X−X bonds differ even within the same compound.

The higher stability of the $s^x p^y$ states, compared with the d configurations, explains why compounds of all the transition metals with a given nonmetal have physical properties which are more similar than the properties of compounds of different transition metals with different nonmetals.

When the nonmetal component is varied along a period, the nature of bonding in crystals varies from the predominantly metallic for metals with high statistical weights of the d^2 configurations to the predominantly covalent for nonmetals with high statistical weights of the sp^3 configurations and the predominantly ionic for nonmetals with high values of the statistical weights of the $s^2 p^6$ configurations. It is important to note that all the main types of bonding which can occur in the lattices of compounds are taken into account in the configurational model in which each type of interaction can be associated with the tendency of the statistical weight of a given type of state to increase.

The Me−Me metallic bonds become more stable, compared with the sp-type Me−X bonds because of increase in the statistical weight of the d^5 states. This explains the following relation-

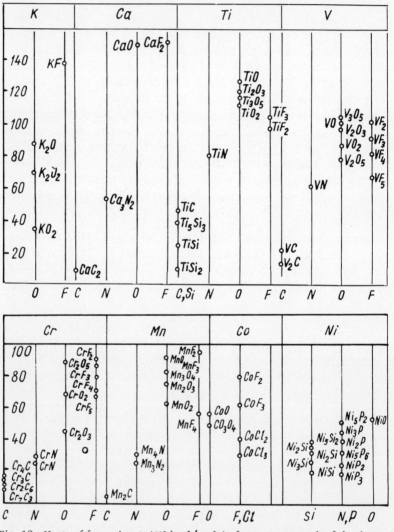

Fig. 12. Heats of formation ($-\Delta H^\circ$ kcal/mole) of some compounds of the d-type transition metals.

TABLE 10. Heats of Formation of Some Refractory
Compounds ($\Delta H°$, kcal/mole)

Metal	Boride	Heat of formation		Carbide	Heat of formation		Nitride	Heat of formation ·	
		total	per B atom		total	per C atom		total	per N atom
Ti	TiB$_2$	72	36	TiC	57.25	57.25	TiN	80.3	80.0
V	VB$_2$	—	—	VC	28.0	28.0	VN	60.0	60.0
Cr	CrB$_2$	>30	>15	Cr$_3$C$_2$	21.01	21.01	CrN	29.5	29.5
Zr	ZrB$_2$	>78	>39	ZrC	44.1	44.1	ZrN	82.2	82.2
Nb	NbB$_2$	>36	>18	NbC	19.0	19.0	NbN	59.0	59.0
Mo	MoB$_2$	23	11,5	Mo$_2$C	4.2	4.2	MoN	17.0	17.0
Hf	HfB$_2$	—	—	HfC	—	—	NiN	88.2	88.2
Ta	TaB$_2$	52	>26	TaC	38.5	38.5	TaN	58	58
W	W$_2$B$_5$	25—45	5—9	WC	8.4	8.4	WN	17	17

ships which govern the heats of formation of a large group of
compounds based on the transition metals (Fig. 12, Table 10).

i) The presence of at least one component with the highest
value of the relevant statistical weight of the stable configurations
reduces the heat of formation. This is supported by the results
obtained for carbides and silicides of the transition metals in
groups I-VIII of the periodic system and by the absence of compounds between these metals and the inert gases.

ii) The heats of formation of compounds are low when the
statistical weight of the stable concentrations in one component
increases at the expense of the statistical weight of the stable
configurations in the other component.

iii) Enrichment of a compound with a metal tends to increase
the heat of formation because of increase in the relevant statistical
weight, in the density of the bonding electrons in the lattice, and
in the contribution of the covalent bonding (this is observed, for
example, for potassium oxides KO_2, K_2O_5, K_2O_2, K_2O; and for titanium oxides TiO_2, Ti_3O_5, Ti_2O_3, and TiO).

iv) A considerable increase in the density of the collective-
state electrons and filling of the bonding levels explains anomalies
in the heat of formation, i.e., the fall of this heat with increasing
concentration of the metal observed for iron oxides, nickel silicide, and vanadium oxides. According to the experimental data
(Table 10), the heat of formation increases with the principal
quantum number of the valence electrons of the metal in the group

IV nitrides and the reverse effect is observed for the group V and VI nitrides. This can only be explained by a gradual change from the p-type nature of the bonding configurations in TiN, ZrN, and HfN, to the d-type nature in CrN, MoN, and WN. A considerable increase in the statistical weight of the d^5 states and a rise of the stability of these states with increasing principal quantum number of the d electrons is responsible for the fall of the heat of formation. These tendencies are much less marked in the case of carbides and borides because of the higher stability of the $s^x py$ configurations which are formed in these compounds.

The heats of evaporation of the transition metals and their compounds can be analyzed in a similar manner [220-221]. Since the evaporation of metals and refractory compounds results in the breaking of the chemical bonds, a measure of the binding energy is provided by the change in the isobaric potential of the evaporation (dissociation) reaction ΔZ_T°. The lower the vapor pressure, the higher the value of ΔZ_T° and the stronger the interaction between atoms (Table 11).

Carbides of the group IV transition metals, for which the Me−C bond is strongest because of the high statistical weight of the sp^3 states, evaporate at compositions which are nearly stoichiometric.

TABLE 11. Isobaric Potentials
ΔZ_T° (cal/g-atom)
of Evaporation of Refractory
Metals and Their
Compounds at 2500°K

Element	Metal	Car- bide	Bor- ide	Ni- tride
Ti	30.5	52.0	64.5	49.0
V	30.8	53.4	—	—
Cr	11.9	21.7	—	—
Zr	64.5	89.5	71.0	61.0
Nb	87.5	93.0	—	—
Mo	70.6	—	—	—
La	32.0	—	36.8	—
Hf	65.5	92.7	—	—
Ta	103 0	97.2	—	2.3
W	113.0	77.5	—	—

In the case of carbides of the group V metals the configuration of a free metal atom is closer to the stable d^5 state and the strongest localization of electrons in the sp^3 configuration is not the most favorable because the redistribution which results in such localization reduces strongly the statistical weight of the d^5 states in the metal and produces a considerable transfer of the remaining d electrons to the collective state. The optimal localization of the electrons in the sp^3 and d states is observed at $x \approx 0.8$ for the compounds $Me^V C_x$ with $x < 1$. At higher values of x more collective-state electrons are formed because of the strong disturbance of the d^5 configurations, whereas at lower values of x the same thing happens because of the considerable reduction in the statistical weight of the sp^3 states. Thus, carbon is lost first from stoichiometric tantalum and niobium carbides because this increases the localization of electrons and reduces the free energy of the system. In the case when $C : Me < 0.8$ the metal is evaporated preferentially and the congruent evaporation occurs at the optimal value of $x \approx 0.8$ [220].

In the case of the group VI carbides the relative enhancement of the Me−Me bonds has an even stronger effect so that the heat of formation of tungsten carbide is about 5 kcal/mole and the vapor pressure above this compound is practically equal to the vapor pressure above graphite. This is evidence of a sharp rise in the statistical weight of the d^5 states in the metal component and of the weak stabilization of the sp^3 configurations in carbon.

Similar relationships would be observed also for borides, nitrides, and oxides but the higher statistical weight of the d^5 states and the greater importance of the Me−Me bonding should have a stronger effect in these compounds (in the case of borides the B−B bonds should also be important). However, the experimental data are not yet sufficient for reliable comparisons and verification of our hypotheses.

The chemical binding forces in the lanthanoids, actinoids, and their refractory compounds are also strongly influenced by the energies and the number of electrons in the d levels. The closer the electron configuration of the atoms of these metals to the f^7 and f^{14} states, the smaller is the fraction of the electrons forming the d configurations and the localized bonds with the neighboring atoms. Therefore, the chemical bonds in the rare-earth metals, actinides, and their compounds with carbon, boron, and nitrogen become weaker on approach of the electron configurations to f^7 and f^{14}. This is manifested most clearly in the evaporation of the rare-earth hexaborides. Since the evaporation breaks the chemical bonds, it is necessary to determine the nature of bonds in the rare-earth hexaborides before considering the evaporation mechanism. In the first approximation, we may assume that a quantitative measure of the strength of the chemical bonds is the heat of atomization, which is equal to the heat of sublimation in the case of metals, and which can be found from investigations similar to those reported in [220] for compounds.

Let us compare the physical and physicochemical properties of the rare-earth metals and the corresponding hexaborides, bearing in mind that the most stable configurations in the f shell are the f^7 and f^{14} states and that the f electrons do not participate directly in the formation of the chemical bonds. This leads us to the following qualitative model of the participation of electrons in the formation of the chemical bonds in the rare-earth hexaborides [222].

When a crystal is formed from atoms of a rare-earth element as a result of random $f \rightarrow d$ transitions, we find that the d electrons (absent in the original free atoms) appear in the solid phase.

Some of the electrons in the 6s shell are transferred to the collective state forming a conduction band and others participate (because of the $s \rightarrow d \rightarrow f$ transitions) in the formation of the stable f^7 and f^{14} electron configurations. The probability of the $s \rightarrow d \rightarrow f$ transitions increases when the configuration of the 4f shell in free atoms approaches a stable state. The probability of the $f \rightarrow d$ transitions then decreases.

This model can be used to estimate the probability of the participation of the f electrons — because of the $f \rightarrow d$ transitions — in the formation of the chemical bonds in the rare-earth metals.

Since europium and ytterbium have the properties of divalent metals, we may assume that the chemical bonds in these elements are solely due to the partition of the 6s electrons and the corresponding heats of sublimation are quantitative measures of the strength of these bonds.

TABLE 12. Contributions of
d Electrons to Chemical
Bonds in Rare-Earth Metals

Metal	Contribution of d electrons	Metal	Contribution of d electrons
La	1.0	Gd	0.98
Ce	1.15	Tb	0.95
Pr	0.77	Dy	0.58
Nd	0.59	Ho	0.56
Pm	—	Er	0.49
Cm	0.12	Tu	0.38
Eu	0	Yb	0
		Lu	1.08

If we assume that the binding energy contributed by the 6s
electrons of metals in the cerium subgroup is equal to the heat
of sublimation of europium and the corresponding energy for members of the ytterbium group is equal to the heat of sublimation of
ytterbium, we can estimate the contribution made by the d electrons to the chemical bonds in the rare-earth metals. This can be
done by subtracting the heat of sublimation of europium or ytterbium from the heat of sublimation of the metal in question. When
the values obtained in this way are related to the energy contributed by the d electrons in lanthanum, we can find the contribution of the d electrons (the probability of the f → d transitions)
to the formation of the chemical bonds. The results of such calculations are presented in Table 12.

TABLE 13. Heats of Dissociation of Rare-Earth
Hexaborides

Compound	Temperature range, °K	ΔH_T°, kcal/mole
La B_6	2045—2300	134.2±2.7
CeB$_6$	1846—2220	123.1±2.8
PrB$_6$	1943—2193	111.5±2.5
	2089—2289	113.0±2.8
NdB$_6$	2133—2355	107.9±2.2
	1983—2290	107.0±1.8
SmB$_6$	2047—2375	102.8±2.3
GdB$_6$	2260—2305	
	2168—2530	127.0±2.8
TbB$_6$	1977—2348	128.7±3.4

The largest relative error in the calculation of the contribution of the d electrons (this error is due to reduction in the energy contributed by electrons with increasing degree of screening of the nuclear charge) does not exceed 20%.

When compounds, such as the rare-earth hexaborides, are formed the behavior of electrons of the rare-earth atoms is basically similar to their behavior in pure metal crystals [152]. However, these electrons now participate also in the formation of bonds between the atoms of boron. The valence electrons of boron participate not only in the formation of bonds between the boron atoms and in the formation of a conduction band but also in the establishment, by the s−d and d− f exchange processes, of stable electron configurations in the 4f shells of the atoms of the metal. The probabilities of the f → d transitions, the tendency of the rare-earth atoms to form stable configurations, and the utilization of the metal electrons in the stabilization of the boron configurations all suggest that the relative strength of bonds in the rare-earth hexaboride series of the cerium group decreases in the middle and at the ends of the periods. This is in agreement with the data on the heats of dissociation (Table 13) and on other thermodynamic properties of these compounds [222].

5. Thermal Properties

The wide use of the transition metals and their compounds in modern technology is a function of their excellent thermal properties. However, there is as yet no theory which would explain the melting or boiling temperatures of these materials. It is only known why the solid−liquid and liquid−vapor phase transitions occur: at sufficiently high temperatures when the entropy factor plays the dominant role in the phase stability, the free energy of a liquid becomes lower than that of a solid and this favors melting of a crystal. The same explanation applies also to the liquid− vapor transition.

It is physically self-evident that, in the case of strongly bound materials, the dominant influence of the entropy on the free energy will be manifested at temperatures much higher than in the case of substances with low binding energies. Since an increase in the statistical weight of the bonding stable configurations increases the strength of binding and reduces the entropy, we may expect a close correlation between the statistical weights of the d^5

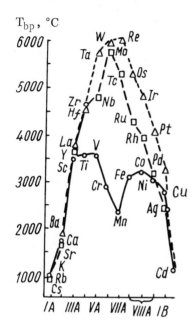

Fig. 13. Boiling points of the d-type transition metals.

states of the pure transition metals and their boiling temperatures. This correlation can be seen in Fig. 13 in which the boiling temperature is plotted as a function of the atomic number.

The general nature of the dependences shown in Fig. 13 and those obtained for other properties controlled by the strength of bonds is in agreement with changes in the statistical weights of the d^5 states (SWASC d^5) listed in Table 2. If an analytic function $T_{bp} = f(P_K)$ (in the zeroth approximation we can use P_5) can be derived, the boiling temperatures T_{bp} of different elements can be calculated by substituting the values of P_5 (which must be calculated for T_{bp}), i.e., the analytic problem should reduce to the solution of the system of equations $T_{bp} = f$ (SWASC d^5) and SWASC $d^5 = f(T)$. The statistical weights of the d^5 states in vanadium, chromium, niobium, and molybdenum are, respectively 63 and 73% ($3d^5$ configurations), and 76 and 84% ($4d^5$ configurations) at $T \approx 0°K$. When the temperature is increased these weights increase for vanadium and niobium but decrease for chromium and molybdenum (curves a, a', and b, b' for Fig. 14). Curves A and A' in Fig. 14 represent the dependences of the boiling temperature on the statistical weights of the d^5 states at that temperature. The strong temperature dependence of the statistical weights of the d^5

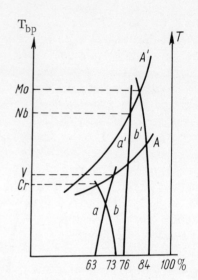

Fig. 14. Schematic representation of the
dependence of the boiling point T_{bp} on
the statistical weight of the stable d^5 configurations (in %) plotted with a suitable allowance for the influence of temperature
on the configuration spectra.

states (SWASC d^5) and the less steeply rising function $T_{bp} = f$ (SWASC d^5) of the 3d metals explain why the boiling point of chromium is lower than that of vanadium, whereas no such anomaly is observed for the 4d metals.

Since the boiling point is a measure of the atomic binding in crystal lattices at high temperatures and the statistical weights of the d^5 states are high even for the VIIIC group metals, the boiling point increases with increasing principal quantum number of the d valence electrons and with increasing stability of the d^5 configurations. On the other hand, the boiling temperature is less for zinc, cadmium, and mercury because of the reduction of the stability of the s and s^2 states (the s-type electrons make the major contribution to the binding in these metals). A superposition of the two effects just mentioned is observed in the copper group. On the one hand the binding forces are influenced by the s-type configurations and, therefore, the boiling point decreases from copper to silver. The further reduction in the stability of s-type configurations and the increase in the stability of the d^5 configurations enhances considerably the contribution of the d electrons to the binding energy and, therefore, the boiling point of gold is higher than that of silver or even that of copper. A similar effect is observed at the beginning of each long period. In the case of K, Rb, and Cs the main contribution to the binding energy is

made by the s electrons (the s → d excitation is weak even at
very high temperatures). Therefore, the boiling point decreases
from potassium to cesium because of decreasing stability of the
s-type configurations. The transition from the alkali to the alkaline-earth metals improves the conditions for the s → d excitation and gives rise to some d^5 states in Ca, Sr, and Ba at high temperatures. The boiling point of strontium is lower than that of
calcium because of the lower stability of the s^2 configurations
(i.e., the main contribution is still that of the s electrons) but the
boiling point of barium is much higher. This can only be explained by the higher statistical weight and the stability of the d^5
states and the larger relative contribution of the d electrons to
atomic bonds in barium, compared with the corresponding parameters of calcium and strontium. In the case of the group IIIA
metals the bonding is dominated by the d electrons and the boiling
point increases along the Sc → Y → La series due to an increase
in the stability of the electron configurations from $3d^5$ to $5d^5$. The
predominance of the d^5 configurations in the bonds in these three
metals is observed only at very high temperatures when the increasing d-type localization raises considerably the statistical
weight of the d^5 states even in the group III metals. At lower temperatures the atoms of these metals tend to retain the d^0 configuration, which favors "expulsion" of the d electrons to the sp
levels and reduction of their participation in the bonding. Therefore, the melting points of the same three metals exhibit a reverse tendency: the decreasing stability of the electron configurations is accompanied by a fall of the melting point along the Sc →
Y → La series (Fig. 15).

In the case of nickel, palladium, and platinum the main contribution to the binding near the melting point is still made by the
d^5 electrons and, therefore, the melting temperature increases in
this group as in the case of the other d metals. The melting point
of chromium is approximately equal to that of vanadium because
the statistical weight of the d^5 states is higher in chromium and
lower in vanadium.

This is also true of the thermal expansion coefficient α, which
is usually measured at much lower temperature (Fig. 16). Obviously, when the statistical weight of the d^5 states and the strength
of the bonds increase, the thermal expansion coefficient should
decrease and, therefore, the lowest value is obtained for chromium.

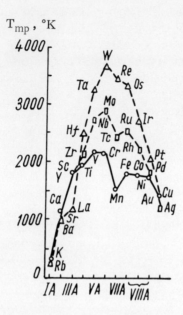

Fig. 15. Melting points of the d-type transition metals.

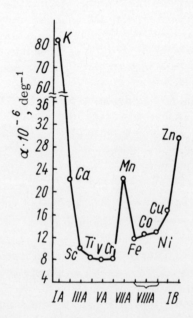

Fig. 16. Thermal expansion coefficients of the 3d-type transition metals.

Manganese has an exceptionally high thermal expansion coefficient and it does not fit the general curve. This anomaly may be due to a considerable fraction of the delocalized antibonding electrons, which are produced by the dissociation of the s^2 configurations.

The behavior of the thermal expansion coefficient also reflects changes in the atomic binding forces above the Debye temperature [223]. A suitable estimate obtained by Gilman [224] shows that the relationship between these parameters can be represented in the form

$$\sigma_c = \frac{3k}{8\alpha V} , \qquad (22)$$

where σ_c is the cohesive strength (stress); V is the atomic volume; and k is the Boltzmann constant.

The thermal properties of more complex systems are also governed by their electron structure, as manifested by the example of compounds between the transition metals and the nonmetals in the second period, beginning from beryllides [225].

The electron configuration of the beryllium atom is $1s^2 2s^2$. When crystalline beryllium and its compounds are formed, a one-electron transition of the s → p type gives rise to the $1s^2 2s 2p$ configuration [147]. The presence of the sp valence electrons favors the formation of relatively strong covalent bonds between the beryllium atoms in the lattice of the metal and in the lattices of its compounds, particularly those formed with other metals. In the latter case, the tendency for the beryllium atoms to form covalent bonds between themselves gives rise to some random loss of electrons from the Me−Be and Me−Me bonds. Therefore, these bonds become weaker.

The presence of the sp states, which tend to become stabilized as the sp^2 and sp^3 configurations, favors the formation of complex structures in the compounds of beryllium, particularly the compounds in which the atoms of the other elements are electron donors. Therefore, in considering the physicochemical nature of the electron and crystal structures of the transition-metal beryllides with partly filled d-electron shells, we must bear in mind the division of electrons in the transition-metal crystals into covalently bound and collective-state electrons. The bonds in beryllides can be represented by the following general formula:

$$\begin{array}{c} \text{Me—Be} \\ \cdot|\cdot \quad \cdot|\cdot \\ \text{Me—Be} \end{array},$$

where the Me − Be bond is mainly metallic with an admixture of covalence governed by the low statistical weight of the sp configurations (this bond is primarily due to the collective-state electrons), the Be $\div$ Be bond is mainly covalent, and the Me $\div$ Me bond has a mixed covalent-metallic nature. Obviously, the relative importance of the various types of bonds in metal beryllides is governed primarily by the ability of the d shell of the metal to accept or contribute electrons in order to form specific configurations.

According to the basic ideas of the configurational localization model, we may expect the III-IV group transition metals to exhibit a tendency to transfer the valence electrons to the beryllium atoms. Moreover, we may expect high statistical weights of the d^0 states and some contribution of the d^5 states in the metal sublattice. A considerable fraction of the d electrons is used in the covalent component of the Me − X bonds. The covalent component of the Me − Me bonds is determined by the statistical weight of the stable d^5 configurations in a crystal. The collective-state electrons are responsible for the stabilization of the sp configurations and, consequently, for the metallic component of the bonding.

Since the statistical weights of the d^5 states in the group IV metals is relatively low, the covalent component of the Me − Me bonds in these metals is very weak.

The increase in the stability of the d configurations in the transition from titanium to zirconium and hafnium and the associated increasing difficulty to form and excite the collective-state d electrons reduces the strength of the Me − Be bonds along the same sequence. The melting points decrease accordingly for the same phases in a group which shares the same metal (Table 14). When the concentration of beryllium is increased, the statistical weight of the sp configurations becomes greater and the Be − Be interaction becomes stronger. This results in a considerable increase of the melting points of the corresponding beryllides, which is illustrated in Table 14 by zirconium and hafnium beryl-

TABLE 14. Melting Points of Beryllides [190]

Phase	T_{mp}, °C	Phase	T_{mp}, °C	Phase	T_{mp}, °C
$TiBe_{12}$	1430	$ZrBe_2$ $ZrBe_5$ Zr_2Be_{17} $ZrBe_{13}$	1200 1530 1750— 1890 1930	$HfBe_2$ Hf_2Be_{17} $HfBe_{12}$	2000 1490 1270
—	—	$NbBe_5$ Nb_2Be_{17} $NbBe_{12}$	1900 1750 1650	Ta_2Be_{17} $TaBe_{12}$	1980 1850
$CrBe_2$	1840	$MoBe_2$ $MoBe_{12}$	1840 1700	WBe_2	2250
—	—	—	—	$ReBe_2$	2000
$FeBe_2$ $FeBe_5$ FeB_{12}	1480 1370 1300	—	—	—	—
$CoBe$	1500	—	—	—	—
$NiBe$	1470	Pd_3Be Pd_2Be Pd_3Be_2 Pd_4Be_3 $Pd_{13}Be_{12}$ $PdBe$	960 1090 1250 1330 1450 1465	—	—

lides. On the other hand, the importance of the Me−Me bonds increases when the concentration of beryllium is reduced and this gives rise to some increase in the melting point of $HfBe_2$ compared with $ZrBe_2$, because of increase in the stability of the configurations which form the Me−Me bonds. A similar increase in the importance of the Me−Me bonds can also be observed in beryllides of the group V and VI transition metals for which the statistical weights of the d^5 states are higher than for the beryllides of the group IV metals. X-ray spectroscopic investigations of the beryllides TiBe, $TiBe_2$, Ti_2Be_{17}, and $TiBe_{12}$ and a comparison with the spectra of pure beryllium [226-227] have demonstrated that the valence electrons of both atoms (titanium and beryllium) are shared in TiBe and $TiBe_2$, which is why the bonds are strongly metallic with a small admixture of the donor−acceptor interaction between the 3d electrons of titanium and the 2d electrons of beryllium. The overlap of the energy bands of the components is considerably weaker in Ti_2Be_{17} and the interaction between beryl-

lium atoms in this compound is much stronger, i.e., strong covalent Be−Be bonds are formed. The importance of the Me−Me
bonds increases in the group V beryllides and, therefore, the transition from vanadium beryllide to tantalum beryllide and from
chromium beryllide to tungsten beryllide is accompanied by an
increase in the melting point because of the rise in the stability
of the d^5 configurations with increasing principal quantum number of the valence electrons. A similar predominance of the
Me−Me interaction is observed also in various phases of the
group V and group VI beryllides: when the concentration of beryllium in a system increases, the statistical weight of the d^5 states
decreases and this reduces the melting point, in contrast to the
group IV beryllides which exhibit the opposite variation as a result of the strong influence of the statistical weight of the sp^3
states on the melting point. Tungsten diberyllide has the highest
melting point (about 2250°C) among the group of compounds we
are considering. It is interesting to note that tungsten and rhenium, which have very stable and not easily disturbed d^5 states,
form phases with the highest concentrations of beryllium ($MeBe_{22}$).
This is evidently due to the fact that the formation of the Be−Be
bonds is more likely than the formation of the Me−Be bonds (because electrons in the stable d^5 configurations are difficult to
excite).

In the transition metals with more than five electrons in the
d shell of free atoms we can expect donor and acceptor properties
(this is due to the large number of possible combinations of the
bonding functions) and a correspondingly large number of compounds with beryllium. This applies particularly to the platinoids. In those cases when the platinoids give up electrons and
the statistical weight of the d^5 states increases, we usually obtain Be-rich phases (as in the case of the group IV-VI transition
metals). This applies, for example, to Ru, Os, and Ir in which
the high stability of the d^5 configurations results in the "expulsion" of the excess electrons to the sp states so that the fraction
of the delocalized electrons is high. If a metal exhibits a tendency to increase the statistical weights of the d^5 states and if it
cannot stabilize effectively the sp configurations of beryllium, we
obtain metal-rich and Be-deficient phases. In such cases the
beryllium atoms cannot easily form covalent bonds. This applies
to palladium and, particularly, to platinum which form phases with
relatively high concentrations of the metal right up to Be/Me =

1/15. However, the same two metals can also form phases with
relatively high concentrations of beryllium (because of the probability of formation of the d^5 configurations) and the melting temperatures of such phases are usually higher than the melting temperatures of the phases with lower concentrations of beryllium.
This happens because the low statistical weight of the d^5 states in
these metals cannot ensure a sufficient strength of the Me−Me
bonds, whereas the Me−Be and Be−Be bonds are weakened by
the reduced concentration of beryllium (reduced statistical weight
of the $s^x p^y$ configurations). Similarly, the melting point of palladium beryllide is somewhat lower than that of nickel beryllide
because the lower stability of the d states ensures a better stabilization of the $s^x p^y$ configurations. The strength of the Me−Me
bonds and the statistical weight of the d^5 states increase on transition from nickel beryllide to cobalt beryllide and this is accompanied by an increase in the melting temperature of CoBe. This
increase in the melting temperature is observed also when the
relative concentration of beryllium is reduced in compounds of
this element with iron because such a reduction increases effectively the statistical weight of the d^5 states in the system and because the influence of these states on the thermal properties of
the beryllides of the elements ranging from group IV to group
VIIIB is not masked by the influence of the statistical weight of
the sp^3 configurations and by the effect of the Me−Be bonds.

Thus, the transition-metal beryllides are characterized by
the formation of covalent bonds between the atoms of the metal
and between the atoms of beryllium and by a simultaneous transfer of some of the electrons to the delocalized state. This gives
rise to weak covalent bonds between the metal and the nonmetal
atoms. These bonds are somewhat stronger in borides, which
are characterized by the formation of structures based on covalent B−B bonds. These bonds between the boron atoms are
stronger than the corresponding bonds between the beryllium
atoms in beryllides because even in its free state boron has a
p electron and further s → p transitions convert the boron structure to the sp^2 configuration with a tendency to form the sp^3 state,
which is most stable in the first half-period.

The sp^3 configurations of boron form the basis of the framework structures of complex borides (MeB_6, MeB_{12}, etc.). The
electron configurations of the boron atoms in diborides are mainly
of the sp^2 type and this is why these compounds have crystal lat-

tices similar to that of graphite, which is also based on the sp^2 configurations and is characterized by the presence of planar networks (network structure). The sp^2 configurations is borides are stabilized if the density of the delocalized electrons is sufficiently high (otherwise they can go over into less stable configurations of the sp, s^2p, and other types). In this case the stabilization reduces to a shift of the $sp^2 \rightleftharpoons sp + p$ dynamic equilibrium to the left in the presence of excess delocalized electrons. Moreover, the collective-state electrons provide bonds between the planar networks of boron atoms. The stabilizing action of the electrons decreases with increasing statistical weight of the d^5 states of the metal component because of increase in the degree of localization. This weakens the bonds within the networks and between them. Borides are characterized by a considerable statistical weight of the $s^x p^y$ configurations which are responsible for the $Me-X$ bonds (these bonds are somewhat stronger than those in beryllides). Consequently, many of the properties of borides, including their thermal stability (Table 15), occupy intermediate positions between the corresponding properties of beryllides and the transition-metal carbides, which are the next class of compounds in this series.

Since the probability of formation of stable d configurations in titanium, zirconium, and hafnium is relatively low, a considerable fraction of the valence electrons of these metals is transferred to carbon when they form carbides. Consequently, the $Me-C$ bonds in the carbides of the group IVA metals are stronger than the $Me-Me$ bonds. The presence of strong $Me-C$ bonds is the reason why titanium, zirconium, and hafnium form only monocar-

TABLE 15. Melting (Dissociation) Points of Some
Refractory Compounds

Boride	T_{mp}, °C	Carbide	T_{mp}, °C	Nitride	T_{mp}, °C
TiB_2	2980	TiC	3147	TiN	2950
VB_2	2400	VC	2810	VN	(2050)
CrB_2	2200	Cr_3C_2	1895	CrN	(1500)
ZrB_2	3040	ZrC	3530	ZrN	2980
NbB_2	3000	NbC	3480	NbN	2300
MoB_2	2100	MoC	2700	MoN	(700)
HfB_2	3250	HfC	3890	HfN	2980
TaB_2	3100	TaC	3880	TaN	3000
W_2B_5	2300	WC	2720	WN	(600)

bides whose melting points are about 1500°C higher than the melting points of the corresponding metals (Table 15).

In the case of the group V carbides the approach of the free-atom configuration to the stable d^5 state reduces the probability of the formation of the sp^3 configurations and increases the statistical weight of the d^5 states of the metal. This means that the strength of the Me−Me bonds increases, i.e., that the share of the Me−Me bonds in the total binding energy becomes higher. The consequence of this is the formation of two types of carbide, MeC and Me_2C, by the group VA metals. These carbides have much narrower homogeneity regions than the titanium, zirconium, and hafnium monocarbides. In the case of the group VA metals the formation of stoichiometric monocarbides is difficult because of the presence of strong Me−Me bonds which tend to reduce the dimensions of the crystal lattice. Evidence of the fall of the statistical weight of the sp^3 configurations in vanadium, niobium, and tantalum carbides, compared with the group IVA metal carbides, is provided by the increase in the melting points of these carbides compared with the melting points of the corresponding metals (this increase is about 1000 deg instead of 1500 deg for the group IV metal carbides [228]).

Carbides of the metals belonging to group VIA form two or three compounds which have very narrow homogeneity regions or none at all. A reduction in the strength of the Me−C bonds in the chromium, molybdenum, and tungsten carbides is manifested by the fact that the melting or dissociation temperatures of these compounds are even lower than the melting temperatures of the corresponding pure metals.

When we compare carbides of the d-type transition metals, we must bear in mind that the stability of the d^5 configurations, which are responsible for the Me−Me bonds in carbides, increases with increasing principal quantum number of the valence electrons of the metal component and this is accompanied by an increase in the melting points of the carbides. For the same reasons the total rate of evaporation decreases, whereas the heat of evaporation, the standard heat of formation, and the energy of atomization increase on transition from titanium carbide to hafnium carbide and from niobium carbide to tantalum carbide. When we go over from the transition-metal carbides to their nitrides, all the parameters which are controlled by the bond energies decrease. This can be understood if we bear in mind that the free-atom configuration approaches the stable s^2p^6 form along the $C-N-O$ series and this is accompanied by isolation and weakening of the $Me-X$ bonds. The covalent component of the $Me-X$ bonds is strongest for carbon ($s^2p^2 \rightarrow sp^3$) and on going over to nitrogen (s^2p^3) and oxygen (s^2p^4) this component is gradually replaced by the ionic component. The covalent component is governed by the statistical weight of the sp^3 configurations of the nonmetal and the degree of stabilization of these configurations by the metal electrons. The ionic component in oxides results from the transfer of the metal electrons to oxygen because of the tendency of this element to attain the s^2p^6 stable configuration.

The nature of bonding varies continuously when the nonmetal component is varied along a period, i.e., oxides retain quite a large covalent component, particularly in the $Me-Me$ bonds. It follows from [98] that only one metal electron is accepted by oxygen in TiO, which is evidence of partial ionic bonding in this compound. However, the tendency to form the s^2p^6 configurations exhibited by oxygen atoms in compounds with the transition metals becomes predominant as we go along a period. Comparative calculations of the energy-band structures of TiO, TiN, and TiC [98] support this conclusion and confirm the weakening of the $Me-X$ bonds, as well as the enhancement of the ionic component, on transition to the oxides.

The hypothesis of a negative charge on the nonmetal in carbides has not yet been confirmed because of lack of reliable experimental data. The most widely held view on the electron structure of refractory compounds is that represented by the "metalloid" theory which postulates the transfer of electrons from the nonmetal atoms to the metal energy band. The recent results [199-202, 439, 442] suggest the opposite effect, namely that the nonmetal atoms accept electrons supplied by the metal atoms and that this effect increases in importance from the middle of the period in question toward oxygen and iodine. The strong ionicity of oxygen and iodine in compounds with the transition metals is the final stage of the enhancement of the ionic component which begins in the middle of the period, where it appears simultaneously with a strong covalent component.

The direct consequence of the above effect is the fall of the melting points on transition from carbides to oxides in all the groups and weakening of their thermal stability (Table 16 [230]).

The unexplained dependence of the melting point on the atomic number of the metal component in the transition-metal oxides can be interpreted by means of the configurational model if we bear in mind that the melting point is governed by the energy

$$\Delta E = \Delta E_1 + \Delta E_2 + \Delta E_3 + \Delta E_4,$$

where ΔE_1 is the increase in the sum of the internal energies of Me^{+n} and O^{-n}, compared with the sum of the energies $Me + O$. This component reduces the melting point whereas ΔE_2, which is the energy of attraction between the Me^{+n} and O^{-n} ions, increases the melting point. The term ΔE_3 represents the energy of the stabilization resulting from the formation of the stable s^2, sp^3, s^2p^6, d^0, d^5, and d^{10} configurations. This component increases ΔE if the Me^{+n} and O^{-n} ions form stable configurations. The destruction of the stable configurations as a result of formation of oxides and the consequent appearance of one of the intermediate configurations, which increase the proportion of the delocalized electrons, reduce the melting points of the oxides. Such destruction increases the energy which is represented by ΔE_4.

It follows from our discussion that the high statistical weight of the empty and the completely filled configurations is responsible for the ionic component of the bonding and the statistical weight of the half-filled configurations governs directly the covalent component. Therefore, we may assume that the high melting point of TiO_2 represents a strong electrostatic attraction in the lattice of this compound because of the high statistical weight of the d^0 and s^2p^6 configurations, whereas extrema of the melting point observed for MnO, Fe_2O_3, and CoO are indications of the high proportion of the ionic bonding as well as a considerable covalent interaction in the $Me-Me$ band. A characteristic feature is the increase of the melting point along the two series $TiO-VO$ and $Ti_2O_3-V_2O_3-Cr_2O_3$ and its decrease along the series $TiO_2-VO_2-CrO_2$. These variations are observed because the capture of some of the metal electrons by the oxygen atoms is favored in the case of titanium and vanadium (and probably in the case of chromium, but the data are insufficient). This is fol-

TABLE 16. Thermal Stability Parameters

Element	Configuration of outer electrons in free atoms	Oxide	Melting point, °C, as thermal stability characteristic	Oxide
Sc	$3d^1\,4s^2$	—	—	Sc_2O_3
Ti	$3d^2\,4s^2$	TiO	1750	Ti_2O_3
V	$3d^3\,4s^2$	VO	2000	V_2O_3
Cr	$3d^5\,4s^1$	CrO	—	Cr_2O_3
Mn	$3d^5\,4s^2$	MnO	1650 (Sublimates and dissociates at 3400)	Mn_2O_3
Fe	$3d^6\,4s^2$	FeO	1368	Fe_2O_3
Co	$3d^7\,4s^2$	CoO	1805—1935	Co_2O_3
Ni	$3d^8\,4s^2$	NiO	1950	Ni_2O_3
Y	$4d^1\,5s^2$	—	—	Y_2O_3
Zr	$4d^2\,5s^2$	—	—	—
Nb	$4d^4\,5s^1$	NbO	1935	Nb_2O_3
Mo	$4d^5\,5s^1$	—	—	Mo_2O_3
Tc	$4d^5\,5s^2$	—	—	—
Ru	$4d^7\,5s^1$	—	—	—
Rh	$4d^8\,5s^1$	RhO	Dissociates at 1127	—
Pd	$4d^{10}5s^0$	PdO	Dissociates above 870	Pd_2O_3
La	$5d^1\,6s^2$	—	—	La_2O_3
Hf	$5d^2\,6s^2$	—	—	—
Ta	$5d^3\,6s^2$	—	—	—
W	$5d^4\,6s^2$	—	—	—
Re	$5d^5\,6s^2$	—	—	Re_2O_3
Os	$5d^6\,6s^2$	OsO	—	—
Ir	$5d^7\,6s^2$	IrO	Unstable	Ir_2O_3
Pt	$5d^9\,6s^1$	PtO	Unstable	—

of d-Type Transition-Metal Oxides

Melting point, °C as thermal stability characteristic	Oxide	Melting point, °C, as thermal stability characteristic	Oxide	Melting point, °C, as thermal stability characteristic
Highly refractory	—	—	—	—
1900	TiO_2	2000	—	—
1970	VO_2	1545	—	—
1990—2265	CrO_2	Dissociates at 427	CrO_3	190
1080		Decomposes at	—	—
(Dissociates above 750)	MnO_2	540-600		
1562	—	—	—	—
(Melts and decomposes)				
Decomposes at 895	—	—	—	—
Thermally unstable	NiO_2	Very unstable	NiO_3	Unstable
2420	—	—	—	—
—	ZrO_2	2700	—	—
1772	NbO_2	2080	—	—
—	MoO_2	Sublimates partly above 1000	MoO_3	795 (sublimates above 650)
	TcO_2	—	TcO_3	—
—	RuO_2	955	—	—
Dissociates at 1115	—	—	—	—
Unstable	PdO_2	Unstable	—	—
2320	—	—	—	—
—	HfO_2	2780	—	—
—	TaO_2	—	—	—
		1270 (sublimates above 1230)	WO_3	1470 under pressure, volatile (sublimates at 1350)
Cannot be obtained in anhydrous form		Unstable	ReO_3	160
		Decomposes at 650	—	—
Decomposes at 400		Partly dissociates above 1100	—	—
—		—	PtO_3	Extremely unstable

lowed by the localization of the remaining electrons (favored by the energy considerations) and the formation of a certain number of the stable bonding configurations. The balance of the components ΔE_1 and ΔE_3, because of increase of ΔE_1, is responsible for the almost constant melting point in the sesquioxide series $Me_2^{IV} O_3 - Me_2^{VI} O_3$. In the case of double oxides the rise of ΔE_1 is responsible for the fall of the melting point right down to chromium. The highest melting point is exhibited by TiO_2 because the constancy of the number of electrons and of configurations favors the formation of the stable d^0 and $s^2 p^6$ states so that the third (negative) term in the total energy of the system increases.

The cobalt and nickel compounds are characterized by low melting points. The sesquioxides and the double oxides of these metals are unstable because the formation of these compounds disturbs the stable d^{10} configurations of the metal component. The high melting points of CoO and NiO are due to the fact that only the s electrons of cobalt or nickel are used to increase the statistical weight of the stable $s^2 p^6$ configurations of oxygen and less energy is needed to detach these electrons from their atoms than to ionize the d electrons.

A characteristic feature of the dependence of the melting point on the composition is the influence of the concentration of oxygen in a given oxide series. In the first long period we observe a fall in the melting point along oxide series in which the formation of the stable $s^2 p^6$ configurations of the oxygen atoms should be accompanied by the destruction of the d^5 configurations of the metal. An example of such a series of oxides is $MnO - Mn_2O_7$. The highest melting point is observed for MnO. This is due to the formation of two stable configurations: the atomic $s^2 p^6$ and the bonding d^5. Further increase in the concentration of oxygen disturbs the d^5 configurations and this is why the melting points of the higher oxides decrease. Similar relationships are observed also for the oxide series

$$VO \rightarrow V_2O_5, \quad CoO \rightarrow Co_2O_3, \quad NiO \rightarrow Ni_2O_3.$$

The decreasing stability of the higher oxides, which is observed, for example, between Sc_2O_3 and Mn_2O_7 (Table 17), is evidently due to a strong increase in the importance of the first term in ΔE (which is not compensated by the ionic attraction when the number of ionized electrons is large) and to a reduction in the energy as a result of the formation of the stable configurations. A different relationship, which is observed in the first long period, is the tendency for the melting point to increase, as observed for the $TiO-TiO_2$ and $FeO-Fe_2O_3$ oxides, i.e., in those cases when the formation of a compound transforms completely the unstable metal and oxygen configurations into the nearest stable atomic and bonding states.

In the second long period the melting point depends on the atomic number in a manner similar to that observed for metals in the first long period. The melting point is practically unaffected when the concentration of oxygen in Rh_2O_3 is increased by comparison with RhO. This behavior should also by observed for RhO_2 in which electrons are redistributed between the stable d^5 and s^2p^6 configurations; in this case there may even be some increase in the thermal stability. The available experimental data for RhO_4 (which represents the next stable state) indicate a low melting point of 27°C, which is in agreement with the relationships observed along the $MnO-Mn_2O_7$ series.

The rise of the stability of the electron configurations with increasing principal quantum number is more pronounced in the third long period than in the second. The melting point of ReO_2 should be the lowest among the rhenium oxides because of increase in the fourth term in ΔE: this is supported by the experimental observations (the double oxide of rhenium is unstable). The stability should increase for Re_2O_7 (297°C) in which the higher stable atomic states of metal and oxygen are formed. The first term of ΔE is fairly large (because of the large charge of the ions) and

TABLE 17. Melting Points of Higher Transition-Metal Oxides

Oxide	T_{mp}, °C	Oxide	T_{mp}, °C	Oxide	T_{mp}, °C
Sc_2O_3	High refractory	Y_2O_3	2420	La_2O_3	2320
TiO_2	2000	ZrO_2	2700	HfO_2	2780
V_2O_5	670	Nb_2O_5	1510	Ta_2O_5	1500—1890
CrO_3	190	MoO_3	795	WO_3	1470 (under pressure)
Mn_2O_7	5.9	Tc_2O_7	120	Re_2O_7	297
	—	RuO_4	27	OsO_4	40

this is why the melting point T_{mp} of Re_2O_3 is relatively low. The monoxide should be highly stable because of the formation of the stable d^5 and s^2p^6 configurations and because of the much higher sum of the terms $\Delta E_3 + \Delta E_4$.

Thus, an increase in the stability of the d^n configurations in an oxide series favors an increase in the oxygen concentration in the most stable oxide and the formation of the stable atomic configurations of the metal and nonmetal because of predominance of the positive term ΔE_3 over the term ΔE_1. In the third long period the energy considerations favor even more strongly the formation of oxides with the atomic configuration of oxygen and the bonding configuration of the metal. This not only stabilizes the ionic $Me-O$ bonds but also increases the covalent component of the $Me-Me$ bonds:

$$CoO\,(1935°\,C) \;\to\; Co_2O_3\,(895°\,C) \;\to\; CoO_2 \;\text{(unstable)},$$
$$RhO\,(1127°\,C) \;\to\; Rh_2O_3\,(1115°\,C) \;\to\; RhO_2 \;\text{(unstable)},$$
$$Ir_2O_3\,(400°\,C) \;\to\; IrO_2\,(1110°\,C).$$

In the three series of oxides given above the formation of the MeO_2 phase corresponds to a redistribution of the electrons between the stable d^5 configurations of the metal and the s^2p^6 states of oxygen, which gives an energy gain ΔE_3 that is largest in the case of Ir ($5d^6$ configuration). In the cobalt and rhenium series of oxides the parallel rise of ΔE_1 peredominates over ΔE_3 and makes the MeO_2 phase unstable. In the iridium series of oxides the condition $\Delta E_1 < \Delta E_3$ is satisfied and the MeO_2 phase has a high dissociation temperature. These relationships, i.e., the fall of the melting point with increasing importance of the term ΔE_1 and the rise of this point on transition to metals with higher principal quantum numbers of the valence electrons, are illustrated clearly in Table 17 which lists the melting points of the higher metal oxides in the three long periods.

When we go over from oxides to sulfides, we find that, in addition to the many relationships governed by the positions of the components in the periods, there are some new features of the dependences which result from changes in the position of the nonmetal component in a given group of the periodic table. The lack of experimental data on this subject makes it difficult to carry out a complete analysis. However, the change in the dependence of the melting point on the atomic number in the group IV sulfides, compared with the corresponding oxides [231], is fully expected and it reflects the fall in the stability of the $s^2 p^6$ configurations of sulfur as a result of the increase in the principal quantum number of the valence electrons of this element.

Analyses of other thermal properties of compounds can be carried out in a similar manner and the relationships governing the variation of these properties in various systems are in good agreement with the configurational model. Since the strongest lattice vibrations are observed along the weakest bonds, we may expect a correlation between the values of the thermal expansion coefficient of the various compounds and the statistical weights of the d^5 stable configurations which govern the Me—Me bonding (this does not apply to some compounds based on the group VI transition metals in which Me—X bonds are the weakest). This is illustrated by the temperature dependence of the relative expansion of refractory carbides plotted in Fig. 17 [232].

The considerable scatter of the results and the different conditions under which thermal properties are usually determined make it difficult to classify and compare them so that theoretical predictions provide a more convenient approach than interpretation of the available data.

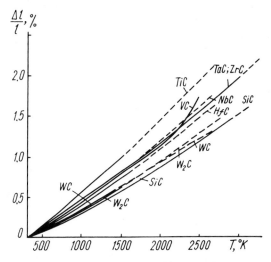

Fig. 17. Thermal expansion coefficients of some carbides of d-type transition metals.

6. Electrical Conduction

The electrical resistivity is very sensitive to the electron structure and therefore it is very desirable to examine the laws governing the resistivity of the transition metals, alloys, and compounds from the point of view of the configurational model.

The electrical resistivity can be regarded as consisting of four principal components [157]:

$$\rho = \rho_f + \rho_i + \rho_m + \rho_e, \tag{23}$$

where ρ_f, ρ_i, ρ_m, and ρ_e represent, respectively, the contributions resulting from the scattering of electrons by thermal vibrations (phonons), impurities, magnetic moments (spin waves), and electrons. The impurity component ρ_i in Eq. (23) is temperature-independent in the case of systems which obey the Matthiessen rule. The magnetic component is constant [$\rho_m = S(S + 1)$, where S is the total spin of an atom] at temperatures above the Curie point and it vanishes for Pauli-type paramagnets. The electronic contribution ρ_e to the resistivity of the transition metals is important only at low temperatures. Therefore, the most interesting component is that due to the scattering of electrons by phonons (ρ_f).

The electrical resistivity of quasifree electrons is governed by the product of the density of carriers and their mobility. The value of the mobility is determined by the intensity of the lattice vibrations (more precisely, by the mean free path of electrons), which is a function of the nature of chemical bonding. The charge carriers in the transition metals, alloys, and compounds are the sp and d electrons. Since the electrons in the stable configurations do not participate in electrical conduction, charge transport by the d electrons is possible if a given substance has intermediate configurations d^k and d^{5+k} ($1 \le k \le 4$) so that the d^1 and d^6 configurations give rise to n-type conduction whereas the d^4 and d^9 configurations give rise to p-type conduction. An allowance for the density of the collective-state electrons and for the strength of chemical bonds in the lattice leads to the following formal expression which is valid provided the temperature is not too low and the carrier density is not too high, so that the last term in Eq. (23) can be ignored:

$$\rho_f = \frac{BT^5}{\Theta^5} \, g\left(\frac{\Theta}{T}\right). \tag{24}$$

Here, g is a dimensionless function defined by the following transcendental expression:

$$g\left(\frac{\Theta}{T}\right) = \int\limits_{0}^{\frac{T}{\Theta}} \frac{x^5 dx}{(e^x - 1)(1 - e^{-x})}.$$ (25)

In the high-temperature approximation (in this case $g \approx \Theta^4/T^4$), we obtain

$$\rho_f \approx \frac{BT}{4\Theta}.$$ (26)

The above formula is practically useless if we wish to predict the nature of the relative change of ρ_f from one substance to another unless we know in advance Θ and n (especially the latter parameter, which is usually deduced from the electrical conductivity and the Hall coefficient).

The constant B is given by an expression which includes the reciprocal of the Debye temperature, the square of the carrier density, and the atomic mass. Therefore, we can isolate the dependences on the carrier density and on the strength of bonds in the equation for the resistivity:

$$\rho_f \approx AT^5 g\left(\frac{\Theta}{T}\right)\frac{1}{n^2\Theta^6} = A'T^5 g\left(\frac{\Theta}{T}\right)\frac{1}{n^2\Theta^6 M}.$$ (27)

The experimental values of the coefficient B for the transition metals are as follows [107]:

Metal	B, $\mu\Omega \cdot cm$	Metal	B, $\mu\Omega \cdot cm$	Metal	B, $\mu\Omega \cdot cm$
Ti	273	Zr	167	Hf	92
V	99	Nb	52	Ta	43
Cr	76	Mo	31	W	25
Fe	59	Ru	48	Os	32
Co	34	Rh	27	Ir	21
Ni	30	Pd	44	Pt	34

If a system exhibits a tendency for high statistical weights of the stable electron configurations and the electrons are localized in the covalent bonds, the values of both parameters (Θ and n) influencing the resistivity are modified. The same tendency also determines the magnitude of the resistivity and the relative change of this property from one substance to another.

The Debye temperature is sensitive to an increase in the covalence bonding in the lattice and it should increase with increasing statistical weight of the d^5 stable configurations, as indicated by the Lindemann relationship (13) between Θ and the melting point of a crystal.

Therefore, the increase in Θ in the first half of a period has a decisive influence on the value of the resistivity which decreases in the same half-period. On the other hand, the change in the statistical weight of the intermediate configurations and in the density of the sp electrons also has an important influence on ρ_f. The values of the factor n^2/A, plotted in Fig. 18, are deduced from the Debye temperatures Θ listed in Table 8 and from the known values of the coefficient B substituted in Eq. (26).

A comparison of Fig. 18 with the dependence of the electronic specific heat on the number of the d electrons (Fig. 11) shows that the dependences plotted in the two figures confirm the importance of the intermediate configurations in the transport of charge in the d-type metals. The general increase of the level of the curve in Fig. 18 (compared with Fig. 11) in the second half-period is the consequence of the contribution of the sp electrons to the transport of charge (in the case of vanadium and chromium these electrons participate effectively in the formation of the s^2 pairs and

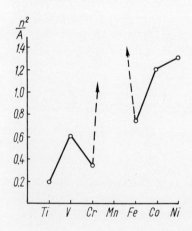

Fig. 18. Electron factor in the resistivity of the 3d-type transition metals.

make a small contribution to conduction, provided $n_{sp} \approx 1$) and of
the higher stability of the intermediate d^{5+k} configurations compared with the d^k states. Moreover, calculations of the type outlined above are inapplicable to manganese because the strong
electron—electron interaction in this metal, resulting from the
·dissociation of the s^2 pairs in the formation of a crystal from free
atoms, enhances the importance of the third term in Eq. (23).

These comments on the nature of the intermediate configurations apply also to the carriers in the transition metals [51]. The
energy stability of the intermediate configurations of the d^k type
increases with increasing k and results in predominance of the p-
type conduction associated with the d^4 and d^9 configurations in the
spectrum of the intermediate states (the d^1 and d^6 configurations,
each of which has one electron in excess of the stable state, are
less favorable compared with d^4 and d^9). Therefore, the Hall effect
measurements usually support p-type conduction in the transition
metals, although undoubtedly there is also some contribution from
the n-type conduction. We can definitely say that the n-type conduction is most important in the case of scandium because the
relatively small number of the localized electrons makes it difficult to achieve not only the d^5 stable configurations but also the
d^4 intermediate configurations (associated with the p-type conduction) due to the fact that this spectrum consists mainly of the
d^1 and d^2 states. The majority of the d electrons is "expelled"
to the sp levels and, therefore, the density of free electrons increases. This n-type contribution becomes less important in the
middle of the period because of a redistribution of the intermediate configurations in such a way as to enhance the proportion of
the d^4 states and because of a reduction in the n-type contribution
to the formation of the stable s^2 pairs in vanadium and chromium.

The n-type contribution to the conduction increases from iron
onwards because of the influence of the collective-state sp electrons, which masks the p-type contribution of the intermediate
spectrum of nickel (in the case of nickel, $n_{sp} \approx 0.6$ and there is
approximately the same number of holes in the d shell; however,
the mobility of the s electrons is higher than that of the holes).

The values of the resistivity of the transition-metal compounds
are similarly determined by the nature of the stable configurations
of the localized valence electrons, the statistical weights and the

stability of these configurations, the proportion of the collective-state electrons and their interaction with the stable configurations.

Since the fraction of the delocalized electrons and the statistical weight of the intermediate configurations in carbides is considerably lower than in the corresponding pure metals, we can expect the high-temperature resistivity ρ to decrease along the series $Me^{IV}C - Me^{V}C - Me^{VI}C$ when the carrier density increases and the statistical weight of the atoms with the sp^3 stable configurations decreases. An increase in the resistivity may be expected for the group VI carbides (especially for WC). The published data on the electrical resistivity are contradictory because of the strong dependence of the electrical resistivity on the presence of impurities and because of wide homogeneity regions in which the properties of carbides can vary considerably. Since it is difficult to obtain reliable information on the electrical resistivity (it would be necessary to prepare exceptionally pure compounds and control carefully their phase composition), the simplest way of investigating the electrical conductivity is to study complex systems and to determine the resistivity of compounds in the homogeneity regions.

Comprehensive investigations of the properties of carbides in their homogeneity regions, beginning from the stoichiometric composition, have been described recently in [233, 235, 237-246].

In analyzing the electrical properties of the carbides it is essential to make an allowance for the second term in Eq. (23) because carbon vacancies can be regarded as impurities. This second term increases in importance with the number of defects in the carbon sublattice and it may dominate the resistivity resulting in a sharp rise of ρ with decreasing value of the ratio C/Me. The results published in [247] and plotted in Figs. 19-20 confirm this hypothesis. In all cases the resistivity rises when the ratio C/Me decreases from 1.

Since the breaking of the Me−C bonds and the formation of carbon vacancies reduces the translational symmetry of a crystal and hinders the appearance of the stable electron configurations, the observed changes in the electron structure can be interpreted as the formation of local regions with higher statistical weights of the stable configurations, which tend to reduce the carrier mobility and make an additional contribution to the resistivity.

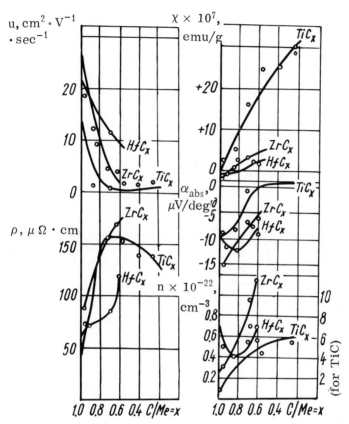

Fig. 19. Some electrical properties of carbides of the group IV
d-type transition metals, measured in the homogeneity regions
of the compounds.

At higher defect concentrations the distances between defects decrease and the interaction between the intermediate electron configurations gives rise to electron transitions which result in the
establishment of a new order in the lattice, i.e., a substructure
is formed. The consequence of these changes is an increase in
the electron mobility and a fall of the electrical resistivity of the
carbides, which can be seen in Fig. 19.

The results of calculations of carrier density in the $Me^{IV}C_x$
carbides, carried out in the single-band approximation on the basis
of the resistivity and the Hall coefficient, are plotted in Fig. 19.
The validity of the single-band approximation is not self-evident

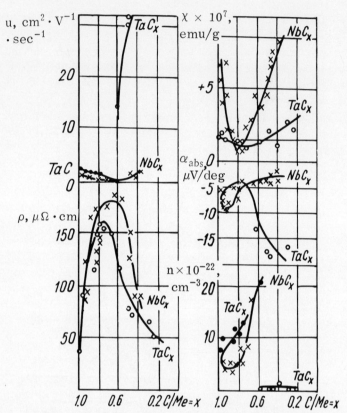

Fig. 20. Some electrical properties of carbides of the group
V d-type transition metals, measured in the homogeneity regions of the compounds.

but the results of measurements of the Hall coefficient indicate
predominance of the n-type component of conduction in all the
carbides, i.e., they confirm quantitatively the correctness of such
calculations. The dependence of the magnetic susceptibility on
the C/Me ratio shows that the covalent type of bonding predominates in the carbides close to the stoichiometric composition and
the importance of this bonding decreases with decreasing concentration of carbon (the relative importance of the metallic bonding
rises). This makes it possible to provide a different and simpler
explanation of the dependence of the carrier density on the ratio
C/Me. Since the covalent component of the $Me-X$ bonds is governed by the statistical weight of the sp^3 stable configurations and

the covalent component of the metallic bonds is determined by the statistical weight of the d^5 states in the spectrum of a transition metal, the rise of the electrical resistivity and of the paramagnetic susceptibility and the fall of the covalent contribution to the total bonding in the lattice can be regarded as due to the reduction in the total statistical weight of the stable configurations as a result of reduction in the concentration of bound carbon in the compound and of increase in the statistical weight of the intermediate configurations.

The results reported in [247] were subsequently refined [442]. In particular, it was found that the electrical resistivity of all the group IV transition-metal carbides rises monotonically (within the homogeneity regions away from the stoichiometric compositions. This rise is accompanied by a corresponding increase in the carrier density, which is supported by the symbatic dependences of the physical properties on the C/Me ratio. The strong influence of the collective-state valence electrons is evidently due to the loss of these electrons from the Me—C bonds which occurs when the concentration of carbon in a carbide is reduced.

The electron structures of the niobium and tantalum, which are characterized by high statistical weights of the stable configurations (~76 and 81%, respectively) in the metallic state, influence strongly the dependences of the electrical properties of the niobium and tantalum carbides on the concentration of bound carbon. Therefore, the tendencies observed for the group IV carbides are manifested even more strongly. The high statistical weight of the d^5 configurations of the metal component has a stronger influence in the group V carbides. The rise of the statistical weight of the d^5 states, compared with the group IV carbides, results in a reduction of the statistical weight of the sp^3 states and weakening of the covalent component of the Me—C bonds. This is due to an increase in the density of the delocalized electrons and enhancement of the importance of the metallic component of the bonds. Therefore the carrier density in niobium and tantalum carbides is one order of magnitude higher than in TiC_x, ZrC_x, and HfC_x.

As in the case of the group IV transition-metal carbides, the electrical resistivity of niobium and tantalum carbides increases away from the stoichiometric composition, reaches a maximum, and then falls. The niobium carbides have two homogeneity regions: one of them represents the MeC phase with the fcc NaCl-type lattice and the other is the Me_2C phase with the hexagonal structure. The rise of the electrical resistivity with decreasing concentration of bound carbon is typical of the NaCl-type phases and it is due to a reduction in the carrier mobility as a result of local fall of the statistical weight of the d^5 and sp^3 states, which is due to partial disturbance of the translational symmetry. However, substructures are observed in niobium and tantalum carbides because of the strong tendency for the statistical weight of the d^5 states of the Nb and Ta atoms to increase at values of the C/Me ratio as high as ~0.8. Therefore, the resistivity decreases rapidly when the ratio C/Me is reduced still further.

The dependence of the carrier density on the C/Me ratio of the group V transition-metal carbides (Fig. 20) is due to the reduction in the number of delocalized electrons at compositions close to stoichiometric as a result of an increase in the statistical weight of the d^5 states and enhancement of the covalent component of the Me − Me bonds accompanied by a weak reduction in the statistical weight of the sp^3 configurations associated with the Me − C bonds. This process continues right up to compositions with the C/Me ratio ~0.8, in which the Me − C bonds are not yet saturated. Further reduction in the relative number of the carbon atoms reduces the statistical weight of the sp^3 configurations, weakens the Me − C bonds (the Me − Me bonds become somewhat stronger because of an increase in the statistical weight of the d^5 states), and increases the number of the delocalized valence electrons. This is supported by theoretical calculations of the Me − Me and Me − X bond energies in the homogeneity regions of refractory compounds (Table 18).

Since the principal quantum number of the d electrons is higher for tantalum than for niobium, the fraction of delocalized electrons in TaC is less than in NbC for any atomic concentrations of carbon. This is due to the localization of electrons either preferentially in the d^5 states (at low values of the C/Me ratio) or in the sp^3 states (at high values of this ratio).

The available experimental data on the electrical properties of vanadium carbide, obtained in its homogeneity region [249], indicate that this carbide behaves similarly to niobium and tantalum

TABLE 18. Bond Energies
in Some Metal-like
Compounds [248]

MeX	Number of electrons in bonds		E_{Me-Me}, kcal	E_{Me-X}, kcal
	Me—Me	Me—X		
$TaC_{1.00}$	1	8	18.7	374
$TaC_{0.75}$	2	6	63.1	280.5
$NbC_{1.00}$	1	8	19.6	357.7
$NbC_{0.75}$	2	6	61.9	269.6
$NbN_{1.00}$	2	6	60.7	281.6
WC	2	8	63	317

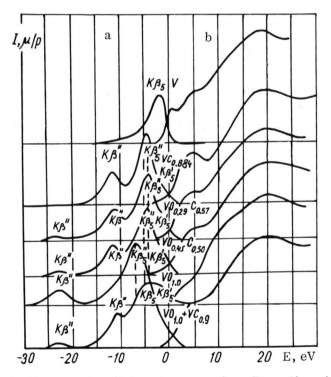

Fig. 21. K emission and absorption spectra of vanadium oxides and carbides.

carbides. The results of x-ray spectroscopic investigations of vanadium carbides and oxides, carried out in their homogeneity regions [250], also show that vanadium compounds behave similarly to the compounds of niobium and tantalum. The x-ray emission and absorption spectra of vanadium carbides and oxides are plotted in Fig. 21. We can see that, when the concentration of the nonmetal becomes higher, the intensity of the $K_{\beta_5}^{I}$ emission band decreases and that of the $K_{\beta_5}^{II}$ band increases. It is usual to attribute the short-wavelength peak $K_{\beta_5}^{I}$ to transitions from the collective-state electron band and to ascribe the long-wavelength peak $K_{\beta_5}^{II}$ to transitions from the bonding electron band. Therefore, the results reported in [250] can be explained if we assume that an increase in the concentration of the nonmetal in the homogeneity regions of vanadium carbides and oxides reduces the fraction of the 3d electrons participating the $Me-Me$ bonds (the intensity of the $K_{\beta_5}^{I}$ band decreases) and increases the number of these electrons in the $Me-X$ bonds (the intensity of the $K_{\beta_5}^{II}$ band increases). The increase in the number of electrons participating in the $Me-X$ bonds is supported also by the increase in the intensity and in the isolation of the long-wavelength peak in the absorption spectra, which is evidently due to the $Me-X$ antibonding states. This is also manifested as an increase in the paramagnetic susceptibility with decreasing concentration of the bound carbon [250]. The same approach can be used in the interpretation of the electrical properties of solid solutions of carbides [251].

The dependences of the electrical resistivity of the solid solutions of the tungsten carbide in the group IV transition-metal carbides (TiC, ZrC, HfC) on their compositions have maxima corresponding to about 30 mol.% WC, with a definite fall on either side of these maxima. The electrical resistivity of the VC−WC and TaC−WC solid solutions increases with increasing concentration of WC, right up to the homogeneity limit (70 mol.% WC). The electrical resistivity of the $NbC-Mo_3C$ solid solutions varies in a similar manner.

Thus, the dependences of the electrical resistivity and the Hall coefficient on the composition of these solid solutions are in full agreement with the property−composition dependences of the metallic binary alloys with a limited solubility in the solid state: a nonlinear rise is observed in the solid solution region and this is followed by an inflection on transition to the two-phase region. A more detailed analysis shows that an increase in the concentration of tungsten in a solid solution reduces the carrier density and their mobility, in accordance with the Nordheim theorem [22] on the conductivity of metallic alloys governed by the scattering of carriers on atoms of different kind. In this case the reduction in the carrier density has a clearer physical origin and it is due to the complex nature of the electron spectra of these systems.

The replacement of the atoms of the group IV metals with the atoms of the group VI metals leads to changes in the nature of the $Me-C$ and $Me-Me$ bonds. The covalent component of the $Me-Me$ bonds increases because of an increase in the statistical weight of the d^5 configurations, compared with the corresponding statistical weights of the group IV metals. Saturation of the covalent component of the $Me-Me$ bonds occurs at the expense of the conduction electrons belonging to the $Me-Me$ energy band. Therefore, the density of these electrons decreases and the electrical resistivity increases. However, since the total number of valence electrons in the group VI transition metals is higher than the number of valence electrons in the group IV metals, the reduction in the density of the conduction electrons occurs only to a certain limit. When the total number of the valence electrons per formula unit reaches ~8.6, the density of the conduction electrons begins to increase because of the saturation of the covalent $Me-Me$ bonds. Since the covalent bonding in the group VI transition metals is stronger than in the group IV and V metals, the covalent component of the $Me-Me$ bonds increases on transition from the group V to the group VI carbides; this corresponds to an increase in the statistical weight of the d^5 states which masks the effect of some fall in the statistical weight of the sp^3 states. This enhancement in the covalent component of the bonds reduces the density of the conduction electrons so that the resistivity of the $VC-WC$, $TaC-WC$, and $NbC-Mo_3C_2$ systems increases throughout the solid-solution regions. The hypothesis of the increase in the statistical weight of the d^5 states as a result of the formation of solid solutions of the group VI transition-metal carbides in the group IV and V carbides and of the influence of this increase on the electrical resistivity and the carrier density is supported by the following observations.

1. The Hall coefficient of the $TiC-WC$, $ZrC-WC$, and $HfC-WC$ solid solutions increases monotonically with increasing con-

centration of WC and becomes positive at the composition corresponding to the electrical resistivity maxima. The Hall coefficient of the VC−WC and TaC−WC solid solutions changes from negative to positive when the concentration of tungsten carbide reaches 30-40%, and the coefficient attains its maximum value at the homogeneity limit (70 mol. % WC). The Hall coefficient of the NbC−Mo_3C_2 system changes from negative to positive at 22 mol.% Mo_3C_2 and attains its maximum value at 70 mol.% Mo_3C_2.

These dependences of the Hall coefficient on the composition indicate a transition from the predominantly n-type to the predominatly p-type conduction which occurs when the concentration of the group V transition-metal carbide is increased. This transition occurs in the region where the Hall coefficient becomes negative. However, it should be stressed that the mechanism of conduction in all such solid solutions is always mixed (n- and p-type). The predominantly p-type conduction mechanism of the solid solutions with the highest values of the resistivity is due to the high values of the statistical weights of the stable electron configurations and the associated predominance of the d^4 and sp^2 intermediate configurations (these configurations are closest to the stable states). The increase in the weight of these intermediate configurations is responsible for the predominantly p-type nature of conduction (translational exchange occurs between these intermediate configurations and the stable states). The contribution of the n-type conduction mechanism, resulting from the translational exchange between the intermediate d^1 configurations and the stable d^0 states, is slight.

2. The above conclusions are in good agreement with measurements of the magnetic susceptibility of the solid solutions under consideration. The results of these measurements demonstrate that the susceptibility decreases with increasing electrical resistivity and the solutions corresponding to the maximum resistivity (minimum carrier density and maximum statistical weight of the stable configurations) have diamagnetic properties.

3. The hypothesis of increase in the covalent bonding in the TiC−WC solid solutions at 20-30 mol.% WC is in good agreement with the position of the maximum in the composition dependence of the microhardness, which is located in the same range of concentrations of WC.

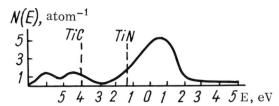

Fig. 22. Density of states in metal-like compounds
(according to Bilz [252]).

4. The results of chemical analysis show that when carbides
of the VI group metals are dissolved in carbides of the IV and V
metals, the concentration of the bound carbon decreases with increasing concentration of the group VI metal. This may be explained by the transfer of electrons from the $Me-C$ to the $Me-Me$
bonds as a result of increase in the statistical weight of the d^5
configurations of the metal component.[*]

These explanations of the nature of the observed composition
dependences of the electrical properties of complex carbides are
in agreement with the Bilz model [252] for the group IV-V carbides: in the region where the number of the valence electrons
per formula unit lies between the values corresponding to the
group IV and the group V carbides, the covalent bonding is stronger, which corresponds to a minimum in the density of states (degenerate gap) in the Bilz model (Fig. 22). However, such an agreement is not observed for the solid solutions of the V-VI group carbides. Since the results obtained can be explained more successfully with the aid of the configurational model, it follows that the
rigid-band model is inapplicable to the transition-metal carbides
having the NaCl-type lattice.

Much less is known about the transition-metal nitrides although the dependences $\rho = f(C/Me)$ obtained for TiN, ZrN, and
HfN indicate that the resistivity depends in a regular manner on
the concentration of the nonmetal component and that this dependence is similar to that obtained for carbides. In the homogeneity
regions the resistivity of the nitrides rises more strongly than
does the resistivity of the carbides because the $Me-Me$ bonds in

_____
[*]An additional contribution to the electrical resistivity of the systems in question is
made by the increase in the number of carbon vacancies.

the nitrides are stronger and the $Me-X$ bonds are weaker. A similar analysis can be carried out also for borides, oxides, and other transition-metal compounds.

The stability of the sp^3 configurations decreases and the statistical weight of the d^5 configurations increases with increasing principal quantum number of the valence electrons of the nonmetal component of a transition-metal compound. The enhancement of the covalent contribution to the $Me-Me$ bonds modifies only slightly the general changes in the electrical properties observed on transition from the $Me^{IV}X$ to the $Me^{VI}X$ compounds (which are similar to the carbides discussed above). Nevertheless, this enhancement increases the number of the silicide and germanide phases compared with the number of carbide phases. The electrical resistivities of the various germanide phases (Table 19) are governed by the same factors as those that apply to carbides in their homogeneity regions: the electrical resistivity decreases when the total statistical weight of the stable configurations in the system increases and the density of the collective-state electrons has a suitable value. For example, a characteristic feature of the group IV transition-metal germanides is a fall in the electrical resistivity with increasing concentration of germanium and the opposite tendency observed for molybdenum germanides. This is due to the fact that the increase in the atomic concentration of germanium and in the statistical weight of the sp^3 states in titanium germanides enhances the strength of the $Me-Ge$ bonds and increases the total statistical weight of the stable configurations,[*] which leads to a fall in the resistivity. Conversely, high values of the statistical weight of the d^5 states and strong $Me-Me$ bonds are more likely to occur in the group VI transition-metal germanides. The highest total statistical weights of the stable configurations and the lowest resistivities should be observed for those systems which have higher concentrations of chromium or molybdenum.

The strong influence of the stable electron configurations on the electrical properties of the rare-earth compounds can be illustrated by considering the rare-earth gallochalcogenides [254].

---

[*] This is supported also by the data on the magnetic susceptibility of the germanides which decreases along the sequences $Ti_5Ge_3 \rightarrow TiGe_3$, $Zr_3Ge \rightarrow Zr_2Ge$, and $Hf_5Ge_3 \rightarrow HfGe_2$, accompanied by the transition of the higher phases to the diamagnetic state.

TABLE 19. Resistivity of Transition-Metal Germanides [253]

Phase	ρ, $\mu\Omega \cdot cm$	Phase	ρ, $\mu\Omega \cdot cm$	Phase	ρ, $\mu\Omega \cdot cm$
Ti_5Ge_3	163.9	Zr_3Ge	142.0	Hf_5Ge_3	127.7
$TiGe$	152.2	Zr_5Ge_3	166.0	$HfGe_2$	58.7
$TiGe_2$	81.7	$ZrGe$	61.8		
		$ZrGe_2$	41.5		
V_3Ge	113.1	Nb_5Ge_3	118.2	Ta_3Ge	331.3
V_5Ge_3	157.1	$NbGe_2$	53.4	Ta_5Ge_3	159.9
$V_{11}Ge_8$	195.0			$TaGe_2$	70.5
Cr_3Ge	30.5	Mo_3Ge	33.5		
Cr_5Ge_3	212.1	Mo_5Ge_3	70.5		
$Cr_{11}Ge_8$	208.0	$Mo_{13}Ge_{23}$	101.7	—	—
$CrGe$	201.0	$MoGe_2$	153.0		
$Cr_{11}Ge_{19}$	273.6				

The results of calculations of the probabilities of electron transitions during the formation of lanthanoid galloselenides indicate random formation of the stable s^2p^6 configuration by two selenium atoms, of the sp^3 configuration by the third atom of selenium and an atom of gallium, and of the $f^{n-1}d^0s^0$ configuration by the lanthanoid atoms. Since all the gallium atoms and one of the selenium atoms in Ga_2Se_3 tend to give rise to high statistical weights of the sp^3 configurations, the interaction between them is basically covalent, which is supported by the tendency of gallium to form the stable sp^3 configurations in sesquichalcogenides. Moreover, the nature of the stable (atomic) configurations formed in this compound indicate that the ionic bonding predominates between the lanthanoid and selenium atoms.

The high statistical weights of the sp^3, s^2p^6, and d^0 configurations are responsible for the semiconducting properties of the lanthanoid gallochalcogenides. This applies particularly to the electrical resistivity of chalcogenides which is typical of semiconductors. The temperature dependence of the resistivity is also of the type encountered in semiconductors as it is due to the dissociation of the stable sp^3 and s^2p^6 configurations and an increase in the statistical weight of the intermediate configurations which can participate in the transport of charge. The changes in the resistivity of gallochalcogenides are governed by their forbidden band widths (0.9-1.8 eV), which are of the order of magnitude expected for stable sp^3 configurations.

If we go over from the $MeSe_2$ compounds to the $Me_2Se_3-Ga_2Se_3$ solid solutions, i.e., if the atoms of the lanthanoid are partly replaced with the atoms of gallium, the forbidden band width decreases and the resistivity becomes smaller. This is due to the fact that the partial replacement of the lanthanoid with the gallium atoms in Me_2Se_3 shifts the electron density maxima toward the cores of the selenium atoms because this reduces the statistical weight of the s^2p^6 configurations (the stability of the s^xp^y configurations of gallium is considerably higher than the stability of the d^n and s^2 configurations of the lanthanoid atoms). Conversely, the addition of a lanthanoid to the lattice of Ga_2Se_3 increases the opportunity for completion of the outer shell of selenium so that it acquires the stable krypton configuration, i.e., the relative statistical weight of the s^2p^6 configurations increases on transition from Ga_2Se_3 to $Ga_2Se_3-Me_2Se_3$. Therefore, the forbidden band width of the solid solutions increases when the atoms of gallium in Ga_2Se_3 are partly replaced with the atoms of a lanthanoid.

The exception to this rule is the solid solution $5Ga_2Se_3-Ce_2Se_3$, whose forbidden band (1.3 eV) is considerably narrower than the corresponding band of pure Ga_2Se_3. This anomaly occurs since the valence electrons of cerium are in the collective state in $Ga_2Se_3-Ce_2Se_3$ (because of the low stability of the f^1 state and the consequent tendency of cerium to increase the proportion of the $4f^0$ stable configurations). The resultant strong electron—electron interaction "loosens" the lattice.

The forbidden band width of the $Ga_2Se_3-Me_2Se_3$ systems decreases along the sequence $Pr-Nd-Sm$ (Fig. 23). According to the configurational model, the statistical weight of the f^7 states of the lanthanoids increases along the same sequence. At the same time the difference between the absolute statistical weights of the stable configurations in the original and the substituent atoms decreases and the localization of the valence electrons in the samarium atoms approaches that observed in the gallium atoms (samarium and gallium are each one electron short for the formation of the stable f^7 and sp^3 configurations, respectively). Consequently, partial replacement of the gallium with the lanthanoid atoms in Ga_2Se_3 (along the sequence $5Ga_2Se_3-Pr_2Se_3 \rightarrow 5Ga_2Se_3-Nd_2Se_3 \rightarrow 5Ga_2Se_3-Sm_2Se_3$) reduces the forbidden band width along the sequence $2.6 \rightarrow 2.5 \rightarrow 2.1$ eV. The last value is

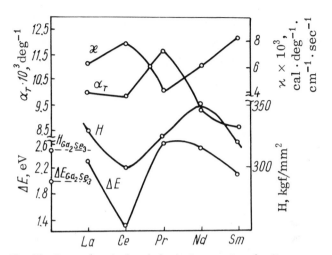

Fig. 23. Some electrical and physical properties of gallochalcogenides of the d-type transition metals.

close to the forbidden band width of Ga_2Se_3 (2 eV) because the
statistical weight of the s^2p^6 states decreases and the statistical
weight of the sp^3 states of selenium increases along the same
sequence. This increases the covalence of the bonding, as supported by an increase in the thermal conductivity and a fall in the
thermal expansion coefficient. The forbidden band width decreases somewhat along the sequence $LaGaSe_3 \rightarrow CeGaSe_3 \rightarrow$
$PrGaSe_3 \rightarrow NdGaSe_3 \rightarrow SmGaSe_3$, in contrast to the solid solutions but in agreement with the increase in the statistical weight
of the f^7 states along the same sequence (the lower limit of the
forbidden band width is set by the considerable contribution of the
f^7 configurations).

The forbidden band width of the rare-earth gallotellurides is
less than that of the galloselenides, which is due to the difference
between their configuration spectra: the reduction in the stability
of the corresponding configurations, which occurs on transition
from selenium to tellurium, tends to increase the overlap of the
wave functions of the electrons, reduce the energy gaps, and lower
the statistical weights of the sp^3 and s^2p^6 configurations.

The available data on the temperature dependence of the re-
sistivity indicate that the differential coefficient $\partial\rho/\partial T$ is related

more closely to the electron structure than the resistivity itself because the influence of impurities on the resistivity is frequently predominant whereas it has little effect on the temperature dependence of the resistivity.

We may assume that the higher the statistical weight of the bonding stable configurations (for the d-type transition metals these are the d^5 configurations), the stronger is the chemical bonding, the smaller the rms amplitude of the lattice vibrations, and the lower the differential temperature coefficient of the resistivity, as revealed by the results of an investigation of the electrical properties of pure metals [255]:

Metal	$\frac{\partial \rho}{\partial T} \cdot 10^2$, $\mu\Omega \cdot cm$	Metal	$\frac{\partial \rho}{\partial T} \cdot 10^2$, $\mu\Omega \cdot cm$	Metal	$\frac{\partial \rho}{\partial T} \cdot 10^2$, $\mu\Omega \cdot cm$
Ti	17.0	Zr	13.6	Hf	13.0
V	6.1	Nb	4.0	Ta	5.1
Cr	4.7	Mo	2.2	W	1.8

When the temperature is raised, the stability of the electron configurations in titanium and its analogs increases whereas the stability of the electron configurations in metals of the chromium group decreases. This should result in weakening of the interatomic bonding and in corresponding deviations of ρ (T) from linearity. Measurements of the temperature dependence of the value of $\partial\rho/\partial T$ of the d-type metals of groups IV-VI [477] have demonstrated that in the case of titanium and its analogs the value of this coefficient decreases with increasing temperature. In the case of group V metals the value of $\partial\rho/\partial T$ remains practically constant in a wide range of temperatures, whereas the corresponding coefficients of tungsten, molybdenum, and chromium increase, particularly in the case of chromium. Similar deviations predicted by the configurational model are observed for yttrium, palladium, and platinum [478].

A study of the published data on the temperature dependence of the electrical resistivity of a large group of metallic alloys has shown that the differential temperature coefficient of the resistivity decreases in a regular manner when configurations with the maximum localization of the valence electrons appear in the electron structure.

Strong Me−C and relatively weak Me−Me bonds are formed in the group IV transition-metal carbides, which are characterized by a high statistical weight of the sp configurations and the low statistical weight of the stable d^5 configurations.

The relatively low carrier density in refractory carbides and the strong variation of this density along the $Me^{IV}C - Me^{V}C - Me^{VI}C$ sequence should have a decisive influence on the variation of $\partial\rho/\partial T$ in all the periods, masking the weaker effect of the reduction in the strength of the bonding. This is supported by the experimental results, which confirm that the differential temperature coefficient of the resistivity decreases from $Me^{IV}C$ to $Me^{VI}C$ as a result of increase in the density of the delocalized electrons because of the fall in the statistical weight of the stable sp^3 configurations and the weakening of the stabilization of these configurations by the electrons of the metal atoms:

Carbide	$\frac{\partial\rho}{\partial T}$, $\mu\Omega \cdot cm$ /deg	Carbide	$\frac{\partial\rho}{\partial T}$, $\mu\Omega \cdot cm$/deg	Carbide	$\frac{\partial\rho}{\partial T}$, $\mu\Omega \cdot cm$/deg
TiC	8.30	ZrC	6.62	HfC	5.07
VC	3.31	NbC	3.50	TaC	2.42
Cr_3C_2	2.32	Mo_2C	1.66	WC	5.55

The slope of the temperature dependence of the resistivity increases strongly only in the case of WC. This is due to the highest stability of the d^5 configurations in this compound, which is responsible for the predominance of the localization of the electrons in the $Me - Me$ bonds. The composition dependence of $\partial\rho/\partial T$ in the homogeneity regions of the carbides differ considerably from the corresponding dependences of the resistivity, which are primarily governed by the contribution of defects in the lattice represented by the second term in Eq. (23).

If we calculate the differential temperature coefficient of the resistivity from Eq. (24), we find that

$$\alpha = \frac{\partial\rho}{\partial T} \approx \frac{5A'T^4}{n^2\Theta^6 M} g\left(\frac{\Theta}{T}\right) - A'T^3 \frac{1}{n^2\Theta^5 M} g'\left(\frac{\Theta}{T}\right). \qquad (28)$$

Hence, the rate of change of α with $x = C/Me$ in a homogeneity region is given by the relationship

$$\frac{\partial\alpha}{\partial x} \approx \frac{2A'T^3}{n^3\Theta^6 M}\left[\left(-5\frac{T}{\Theta}g + g'\right)\Theta\frac{\partial n}{\partial x} + \left(-15\frac{T}{\Theta}g + 5g' - \frac{\Theta}{2T}g''\right)n\frac{\partial\Theta}{\partial x}\right]. \qquad (29)$$

Since, $g'' < g' < g$, the polynomials in parentheses are all of the same order of magnitude. It follows from Eq. (29) that change in the carrier density and in the bulk modulus of the lattice (governed by the Debye temperature) influence strongly the differential temperature coefficient of the resistivity. An analysis of the relative importance of these factors can be carried out on the basis of suitable information on their composition dependences in a homogeneity region. An allowance for the possibility of the d^5 and the sp^3 localization and the formation of the $Me-Me$ and $Me-X$ covalent bonds shows that the carrier densities obey the following relationships (Figs. 19 and 20):

$$n_{TiC_x} \ll n_{VC_x},$$
$$n_{ZrC_x} \ll n_{NbC_x}, \tag{30}$$
$$n_{HfC_x} \ll n_{TaC_x}.$$

Moreover, the higher values of the statistical weight of the sp^3 states in the titanium, zirconium, and hafnium carbides, compared with the values typical of the corresponding $Me\,C$ phases, affect the bulk modulus of the lattice. The available data indicate that the Debye temperatures obey

$$\Theta_{TiC_x} > \Theta_{VC_x},$$
$$\Theta_{ZrC_x} > \Theta_{NbC_x}, \tag{31}$$
$$\Theta_{HfC_x} > \Theta_{TaC_x}.$$

It follows from Eqs. (30), (31), and (32) that the differential temperature coefficient of the resistivity of the group IV transition-metal carbides is governed primarily by the carrier density (an increase in the density in the homogeneity regions of these compounds masks the influence of the reduction in the rigidity of the lattice), whereas the values of $\partial\rho/\partial T$ of the group V carbides are affected more strongly by the bulk modulus of the lattice.

Thus, we find that the function $\partial\rho/\partial T = f(C/Me)$ is satisfactorily "anticorrelated" with the increase in the carrier density resulting from the decrease in the ratio $x = C/Me$ of TiC_x and ZrC_x and with the more complex variation of the carrier density in HfC_x. Conversely, in the case of $Me^V C$ carbides we should observe a fairly close correlation with the strength of bonds in the lattice. Therefore, the differential thermal coefficient of the re-

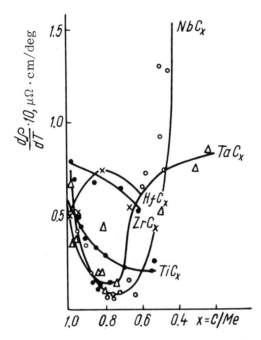

Fig. 24. Dependences of the temperature coeffi-
cient of the resistivity $\partial \rho / \partial T$ on the carbon content in some transition-metal carbides in their
homogeneity regions.

sistivity has its smallest value at C/Me $\approx$ 0.8 (Fig. 24), which
corresponds to an increase in the combined statistical weight of
the d^5 and sp^3 configurations and is in agreement with changes in
the other properties of carbides in the homogeneity regions.

An analysis of the recent data on the thermal coefficients of
the resistivity of complex carbides [255] shows that the temperature dependence of the resistivity of the TiC−WC system is in
agreement with the metallic nature of all the compositions, with
the exception of the $(Ti_{0.8}W_{0.2})C_{0.96}$ and $(Ti_{0.7}W_{0.3})C_{0.96}$ phases
which have negative values of $\partial \rho / \partial T$ (the negative value is evidence of semimetallic nature). The values of $\partial \rho / \partial T$ of the ZrC−
WC and HfC−WC systems also pass through minima correspond-
ing to 30-40 mol.% of WC. The $(Hf_{0.6}W_{0.4})C_{0.92}$ phase is characterized by a negative $\partial \rho / \partial T$, which indicates that the electron
scattering mechanism is close to that encountered in semicon-
ductors.

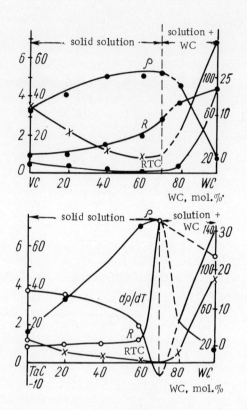

Relative temperature coefficient (RTC) $\partial \rho / \cdot \partial T \times 10^3$, deg^{-1}

$\partial \rho / \partial T \times 10^3$, $\mu\Omega \cdot$ cm/deg

ρ, $\mu\Omega \cdot$ cm

$R \times 10^4$, cm^3/C

Fig. 25. Dependences of the electrical properties of VC—WC and TaC—WC solid solutions on their composition.

A characteristic feature of the composition dependences of the temperature coefficient of the resistivity of the TiC—WC, ZrC—WC, and HfC—WC systems is a minimum in the same range of concentrations in which minima (or negative values) of the magnetic susceptibility and maxima of the electrical resistivity, microhardness, and the Hall coefficient are observed (Fig. 25).

The dependences shown in Fig. 25 have their origin in the considerable rise in the statistical weight of the stable d^5 configurations at the optimal concentrations of WC in the solid solutions, which is primarily due to the destruction of the intermediate configurations and the slight weakening of the stability and the statistical weight of the sp^3 configurations. The net effect of the increase in the localization is an enhancement of the strength of bonds in the lattice and a weakening of the electron—phonon scattering.

The temperature coefficient of the resistivity of the VC—WC solid solutions decreases with increasing concentration of tungsten right up to the limit of the homogeneity region. This is also observed for the TaC—WC system but in this case the temperature coefficient of the resistivity becomes negative for the $(Ta_{0.3} \cdot W_{0.7})C_{0.9}$ composition.

The presence of directed covalent bonds has the same effect as in the $Me^{IV}C$—WC solid solutions: it reduces the amplitude of the thermal vibrations and of the slope of the straight lines $\rho(T)$, as observed experimentally. The solid solutions with the highest contribution of the covalent bonding (deduced from measurements of the magnetic susceptibility, Hall coefficient, and other properties) have small values of $\partial\rho/\partial T$.

Some of the complex solid solutions of the carbides have negative values of the temperature coefficient of the resistivity, minimal but relatively high (10^{21}-10^{22} cm^{-3}) values of the carrier density, and low values of the electrical resistivity and of the thermoelectric power (this is typical of metals). Therefore, these solid solutions should be regarded as semimetals. At sufficiently low temperatures and in the case of weak scattering of carriers by phonons, the low-temperature mechanisms of carrier scattering typical of semiconductors may be encountered in such semimetals. This is why the temperature dependence of the resistivity of some of the carbide solid solutions resembles that observed for semiconductors.

If we now consider some of the individual compounds, we find that the differential values of the temperature coefficient of the resistivity of borides are numerically similar and are higher for the group V and VI borides than for the corresponding carbides. This is basically in agreement with the nature of changes in the statistical weights of the d^5 and the sp^3 configurations and in the bulk moduli of the compounds in question. However, the differential temperature coefficients of the resistivity of the group IV borides are considerably lower than the corresponding coefficients of the group IV carbides and the bulk moduli of the former are much lower. In this case the value of $\partial \rho / \partial T$ is affected much more strongly by that factor which results in a reduction of the temperature coefficient with increasing number of carbon vacancies in the lattice of titanium carbide. The density of the collective-state electrons in TiB_2 is higher than in TiC, because in the latter compound this density is reduced by the high value of the statistical weight of the sp^3 configurations.

The electrical properties of disilicides are in many respects similar to the properties of diborides. This is due to the special features of their configuration spectra: the statistical weights of the sp^3 configurations in borides are lower than in carbides, whereas the stability of these configurations in silicides is lower than that in carbides. For this reason weakening of the $Me-X$ bonds (and the related fall in the strength of the lattices of silicides) is accompanied, as in the case of borides, by an increase in the strength of the $X-X$ bonds (these bonds are particularly responsible for the very high stability of the MeB_2 and $MeSi_2$ phases) and of the $Me-Me$ bonds which exert an important influence on the electrical properties. The stronger $Me-Me$ bonds in disilicides make the behavior of $\partial \rho / \partial T$ similar to that observed for pure metals. The lower stability of the sp^3 configurations in disilicides makes the number of carriers per atom of metal in these compounds higher than the corresponding numbers for carbides and borides. On the other hand, the bulk moduli of disilicides are lower than those of carbides and borides. Calculations which yield expressions similar to those in Eqs. (30) and (31) show that the rigidity of the lattice has the strongest influence on the differential temperature coefficients of the resistivity of silicides. Therefore, the values of $\partial \rho / \partial T$ of the MeSi phases vary, like those of borides, with the rigidity of the lattice and with the statistical weight of the d^5 configurations of the metal.

7. Emission Properties

It follows from general considerations that the electron structure of a solid, particularly the structure of the surface layers, determines the emission of electrons. Therefore, the work function of elements has been successfully correlated with their atomic volume [257-259], compressibility [260, 261], first ionization potential [262], crystal lattice energy [263-264], surface tension

[230, 265-266], electronegativity [264], zero-charge potential, and atomic radius [267].

The work function is an energy characteristic of a material because it is determined by the potential energy of the binding of electrons in a solid. Since the d energy bands are narrow and since they overlap the s and p bands, the valence electrons in the transition metals exhibit behavior which is intermediate between the Heitler−London and the Bloch electrons, and this is reflected in the hypothesis of their partial localization. Therefore, an increase in the density of the delocalized electrons implies weaker binding of each of the electrons to the lattice. In the case of the elements belonging to the first transition row between scandium and chromium and in the case of their analogs in the same group, the proportion of the delocalized sp electrons decreases with increasing value of the statistical weight of the d^5 configurations and these electrons become localized in the stable configurations. We may expect this to result in a monotonic decrease in the work

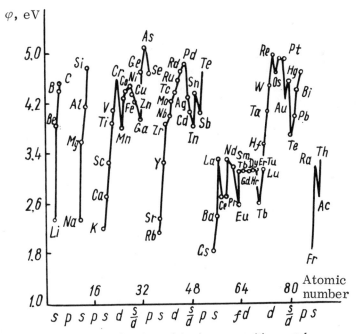

Fig. 26. Work functions of the d-type transition metals.

function, which is confirmed by the experimental results (Fig. 26). Chromium not only has a high value of the statistical weight of the d^5 states but also has about one s electron per atom. Therefore, stable s^2 pairs form in the solid phase and fewer electrons are transferred to the collective state. Consequently, the work function of chromium is the highest among the metals belonging to the first half-period. The low work function of manganese may be attributed to the considerable density of the collective-state s electrons produced by dissociation of the s^2 configurations on transition from free atoms to a solid. These s electrons can escape easily from the solid under thermionic emission conditions.

We shall not consider the full details of the rising part of the curve in Fig. 26, except to state the general conclusion that the work function of the transition metals increases with increasing weight of the d^5 states, i.e., with increasing fraction of the electrons localized in energetically stable states.

In the case of metals with $n_d > 5$, i.e., the transition metals of group VIII, we must bear in mind that the stable states in these metals are d^5 and d^{10} and that the higher stability of the d^{10} state compared with the d^0 state implies that the proportion of the valence electrons going over to the collective state decreases. Consequently, we may expect a rise in the work function with increasing number of electrons in free atoms. It is clear from Fig. 26 that when the principal quantum number increases, i.e., when the energy stability of the d^n configurations becomes higher, the work function increases in a regular manner. The general rise of the work function with increasing number of the d electrons is less marked for the metals with $n_d > 5$ than for those with $n_d < 5$, which is due to the smaller difference between the energy stabilities of the d^5 and d^{10} states than that between the stabilities of the d^0 and d^5 states. Thus, the work function of the metals in the second half-period rises with increasing statistical weight of the stable d^{10} configurations. Hence, we can draw the important practical conclusion that in order to reduce the work function we must transform a substance in such a way as to reduce the statistical weights of the d^5 and d^{10} configurations and increase as much as possible the density of the collective-state electrons.

The curves plotted in Fig. 26 illustrate clearly the functional relationship between the work function and the electron structure.

Since the surface of a crystal differs from its bulk by the lower values of the statistical weights of all the stable configurations, it follows that the experimentally determined values of the work function are lower than those which would be obtained for thermionic emission from the bulk of a crystal. The only question remaining is the extent to which the statistical weight of a given stable configuration is reduced on the surface of a crystal compared with its bulk value. The available experimental data indicate that, in the first approximation, the change in the stability is the same for all configurations.

All the results quoted so far refer to the "integrated" work function of polycrystalline samples. In the case of single crystals, the work function differs somewhat from one face to another. This is due to the fact that the statistical weights of the stable configurations are different for different faces and directions and for different polymorphic modifications of the same substance.

It is physically self-evident that, in the presence of a spectrum of configurations, the electron exchange between the states with definite stabilities must take place along certain preferred directions. For this reason, there are directions with higher and lower values of the statistical weights of the stable configurations and also directions with minimal and maximal localization of the electrons.[*] The faces with the minimal localization will have the lowest and those with the maximal localization the highest work functions. When the statistical weights of the stable configurations increase, the contrast between the configuration spectra for different directions in a crystal should increase and, therefore, the difference between the work functions of different faces should also become greater. If the general level of the statistical weights of the stable configurations is low, the difference between the work functions of different faces should decrease and the energy states of the faces should, to some extent, be leveled. In the hypothetical case of complete delocalization of the valence electrons, the difference between the work functions of

---

[*] In this respect, the results that follow from the configurational model are in many respects similar to those obtained from the band theory, which predicts that the dependences $E(\mathbf{k})$ have different forms for different directions of the wave vector $\mathbf{k}$ in the reciprocal space. It follows that the population of the d energy band depends on the direction in a single crystal.

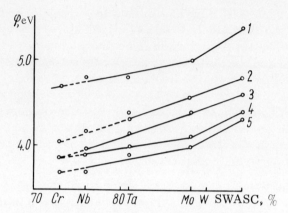

Fig. 27. Dependences of the work functions of various faces of single crystals of the d-type transition metals on the statistical weight (SWASC) of the d^5 states: 1) $\{110\}$; 2) $\{112\}$; 3) $\{100\}$; 4) $\{111\}$; 5) $\{116\}$.

the various faces should vanish (under these artificial conditions, the faces could not form at all).

Experimental data on the work functions of different faces of single crystals are very scarce. It is reported in [268] that the dependences of the work functions of the transition metals on the statistical weights of the stable configurations (Fig. 27) vary in a similar manner for different faces. Let us consider the results obtained by extrapolating the dependences of Fig. 27 to lower values of the statistical weights of the atomic stable configurations (SWASC). If this is done for the $\{112\}$ faces of niobium and a range of faces of chromium, the following values of the work function are obtained:

for Nb: $\varphi\{112\}$ ~ 4.20 eV;

for Cr: $\varphi\{116\}$ ~ 3.75 eV;
$\varphi\{111\}$ ~ 3.88 eV;
$\varphi\{100\}$ ~ 3.88 eV;
$\varphi\{112\}$ ~ 4.05 eV;
$\varphi\{110\}$ ~ 4.70 eV.

An analysis [268] of the difference between the work functions of two faces, $\Delta\varphi = \varphi\{110\} - \varphi\{111\}$, shows that there is a definite tendency (Fig. 28), whose origin is discussed above. A detailed

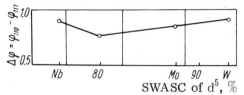

Fig. 28. Dependence of the difference between the work functions of two different faces of single crystals of d-type transition metals on the statistical weight (SWASC) of the d^5 states.

study [268] shows that the work functions of specific faces of single crystals vary in the same manner as the work functions of polycrystalline samples, whereas the difference between the work functions of the faces of a single crystal of the same material are governed by the differences between the statistical weights of the stable configurations typical of each face.

These ideas can be used to interpret more reliably the emission properties of various refractory compounds formed by transition metals with nonmetals. In such compounds, the metal and the nonmetal atoms tend to form stable electron states of one kind or another. Figure 29 shows the work functions of diborides of the d-type transition metals belonging to groups IV-VI. It is worth noting that the work function of the diborides of the 3d and 4d metals decreases monotonically with increasing statistical weight of the d^5 states of the metal component. In the third long period, the variation of the work function is more complex: it decreases on transition from HfB_2 to TaB_2 and then it increases for W_2B_5. Similar rises and falls are observed if we go from one group to another. For example, in the case of borides of the group IV metals, the work function decreases with increasing atomic num-

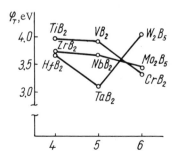

Fig. 29. Dependences of the work functions of diborides of the d^5 transition metals on the number of (s + d) electrons in the metal.

ber of the metal, whereas in the case of borides of the group V metals, the work function shows a tendency to increase (tantalum boride), which is replaced by a monotonic rise in the case of the group VI borides.

The importance of providing an explanation of these tendencies is self-evident but the purely "surface" approach to the nature of the work function fails to give even a correct formulation of the problem. If we use the configurational model, the observed variations can be shown to be correlated closely to other properties depending on the electron structure of a given class of substances, and these variations can be explained as follows. The group IV borides are characterized by high values of the statistical weights of the sp^2 states, which decrease when we go over to the group V borides. In the first long period, the statistical weight of the sp^2 states continues to decrease as we move toward chromium boride. This happens because the relatively low stability of the $3d^5$ configurations prevents effective localization of the d electrons in the Me−Me bonds, and the low values of the weight of the sp^2 states weaken the localization of the electrons in the Me−X bonds and reduce the potential energy of the binding of these electrons. In the third long period, the high stability of the d^5 configurations, which is typical of tungsten, results in the strong localization of the electrons in the d^5 configurations and in the B−B bonds. Consequently, the work function increases strongly. The same effect (though less pronounced) should be observed also for Mo_2B_5, which is confirmed by an analysis of changes in other properties of this series of borides. Therefore the results reported for MoB_2 in [269] should be regarded as overestimates.[*]

The nature of the variation of the work function in a given group of the metal component can be explained by an increase in the importance of the Me−Me bonds and of the degree of localization of the electrons in these bonds. For example, in the case of the group IV diborides, the work function has its highest value for TiB_2 and decreases to ZrB_2 and HfB_2, because of the corresponding decrease in the values of the statistical weights of the sp^2 configurations and the stabilization of these configurations by the metal electrons. In group VI, the situation is reversed because of the very strong localization of electrons in the d^5 configurations. The work function should increase with increasing

---

[*] The values of the work functions of W_2B_5 and Mo_2B_5 were taken from [270].

statistical weight of the d^5 states and increasing stability of these states: this is indeed observed. In the vanadium group, some competition between the two effects takes place: the transition from VB_2 to NbB_2 tends to reduce the statistical weight of the sp^2 states in the system and therefore the work function decreases somewhat. However, further increase in the statistical weight and the stability of the d^5 configurations in TaB_2 weakens the localization of the electrons in the $Me-B$ bonds. However, this effect masks the influence of some increase in the strength of the $Me-Me$ bonds and reduces the work function.

Until recently, it was not possible to explain the variations observed in the work function of carbides of the transition metals belonging to groups IV-VI. According to the ideas outlined above, this class of compounds should obey relationships basically similar to those observed for borides, i.e., the work function should decrease in the transition from $Me^{IV}C$ to Me^VC and it may increase for carbides of the group VI metals (since the strength of the $Me-Me$ bonds in carbides is less than in borides, the last effect should be weaker in carbides). The numerous published data on carbides are usually not comparable because of the different methods used to measure the degree of contamination of the samples (which was generally not controlled), the porosity, and the different temperatures at which the measurements were carried out. A gradual review is being made of the work functions obtained earlier for carbides of the group IV-VI metals and the results obtained tend to fit the predicted dependences. This is supported by the results reported in [239] on the variation of the effective work function of carbides within their homogeneity regions, because the existence of these regions is an additional cause of the disagreement between the published results. Papers on the thermionic emission of carbides fail to give their composition. This is an important point because it is now known [239] that the work function of a given carbide can vary considerably within this homogeneity region (Fig. 30). This variation is not very large in the case of niobium and tantalum monocarbides but it is very strong in the case of titanium and zirconium monocarbides because of the special features of their configuration spectra.

When titanium carbide is formed, the delocalized fraction of the valence electrons of the titanium atoms is transferred to the carbon atoms. This results in the stabilization of the sp^3 con-

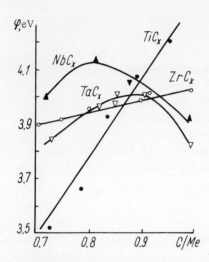

Fig. 30. Dependences of the work functions of refractory carbides on the C/Me ratio (homogeneity regions at 1800°K).

figurations and in the transfer of some of these electrons to the covalent components of the Me−C bonds. The statistical weight of the sp^3 configurations increases continuously with increasing concentration of carbon within a homogeneity region. A linear rise of the work function is evidence of an increase in the degree to which electrons are bound as a result of the Me−C interaction. The Ti−Ti covalent bonds are not very important because the statistical weight of the d^5 configurations associated with the metal is small. This is also observed for ZrC but, because of the higher stability of the d^5 configurations in zirconium (compared with titanium), the statistical weight of the sp^3 states and the work function decrease more slowly within the homogeneity region of ZrC. The Me−Me bonds in this component are somewhat stronger than in titanium carbide.

The nature of the dependence $\varphi = f(C/Me)$ changes considerably when we go over to carbides of the group V metals because of the sharp rise in the d localization and the increase in the strength of the Me−Me bonds. The work function φ first rises with increasing concentration of carbon (as in the case of the $Me^{IV}C$ carbides) because of the stronger sp^3 localization. This is accompanied by a gradual weakening of the Me−Me bonds. Initially, this is not of great importance but eventually it predominates over the increase in the statistical weights and the stability of the sp^3 configurations. A further increase in the concentration of carbon, beginning from C/Me ≈ 0.8 in the case of NbC and from C/Me ≈ 0.9 for TaC, results in a considerable weakening of the binding of the electrons in the lattice and a consequent reduction of the work function. In the case of TaC, the work function maximum lies at higher concentrations of carbon than in the case of NbC because the stability of the d^5 configurations and the localization of the electrons is higher for tantalum than for niobium and the strength of the covalent Ta−Ta bonds varies more slowly with the concentration of carbon than does the strength of the Nb−Nb bonds. The nonlinear dependences of the work function of niobium and tantalum carbides on the concentration of carbon are well correlated with the variations, in the homogeneity regions, of other physical properties such as the lattice parameters, electrical resistivity, Hall coefficient [241], microhardness [242], superconducting transition temperature [243], density [244], thermal expansion coefficient [245], heat of formation [246], and magnetic susceptibility [238].

In the stoichiometric region, the work functions falls along the sequence TiC → ZrC → NbC → TaC because of the reduction in the statistical weights of the sp^3 configurations and the weakening of the stabilization of these configurations by the metal electrons. Consequently, it is easier to excite electrons under thermionic emission conditions. The published values of the work function of the group VI carbides are highly contradictory. The results discussed above suggest that the work function in the first and second long periods continues to decrease from the Me^VC phases to the carbides of chromium and molybdenum because of the further decline of the statistical weights of the sp^3 configurations which is not compensated by an increase in the d localization. A comprehensive comparison of the properties of these compounds suggest that tungsten carbide is characterized by the largest differences between the sublattices and by the strongest localization of the electrons in the Me−Me and C−C bonds. These features of the configuration spectrum of tungsten carbide are the consequence of the highest values of the statistical weights of the d^5 states and of the stability of these states in the metal component, and of the analogous extremal characteristics of the configuration spectra in the carbon sublattice. In the final analysis, this leads to a considerable increase in the work function. According to recent investigations [271], the work function of tungsten carbide is about 4.3 eV.

The development of the ideas on the stability of the various d and f configurations makes it possible to provide a clearer interpretation of the variations in the work function of the rare-earth metals and their compounds [181]. In the first group of lanthanoids, whose free atoms have less than seven f electrons, the statistical weight of the f^7 configurations increases as the number of f electrons increases. Since the s electrons dominate emission phenomena and since these electrons are partly transferred to the f states, which results in an increase in the statistical weight of the f^7 configurations and a consequent increase in the binding energy of these electrons in a crystal, it follows that we can expect the work function of these metals to increase in a regular manner on approaching the middle of the period.

The work function of europium is considerably lower than that of samarium because the f shell of europium represents the stable f^7 state and practically all the s electrons participate in thermionic emission since they are not needed to form the stable f state.

In the case of gadolinium ($f^7d^1s^2$), we have one electron in the d state. The tendency of the d state to stabilize gives rise to the $6s \rightarrow 5d$ transitions, which reduce the number of s electrons (these are the electrons that can participate in thermionic emission), increase the binding of these electrons to the crystal, and raise the work function of gadolinium compared with that of europium (Fig. 31).

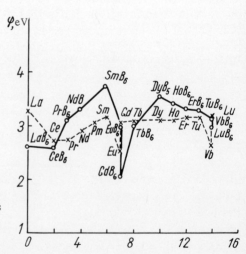

Fig. 31. Dependences of the work functions of f-type transition metals and their hexaborides on the number of f electrons.

The f-type elements with more than seven f electrons per atom form, in the metallic state, stable f^7 and f^{14} configurations. Since the f^7 and f^{14} configurations differ less in energy than the f^7 and f^0 configurations, the rise of the work function with increasing statistical weight of the f^{14} configurations is weaker but the ratio of the work functions of ytterbium and lutetium can be explained in the same way as has just been done for europium and gadolinium. In the formation of compounds such as the rare-metal borides, the atoms of boron, which have the s^2p valence electron configurations in the free state, acquire the more stable sp^2 configuration by the s → p transitions. The sp^2 configuration then transforms into an even more stable configuration of the sp^3 type. This second transformation enhances the localization of electrons and this usually tends to increase the work function of hexaborides compared with that of the corresponding pure metals (Fig. 31).

It should be pointed out that the excess of the work function of a hexaboride over that of the corresponding pure metal generally increases with increasing number of the f electrons in the free atoms, which is evidence of the increasing tendency to complete the partly filled f shell and to raise the statistical weight of the stable f-type configurations.

The atoms of those lanthanoids which have electrons in the d shell exhibit the s → d transitions, which result in some growth in the work function. When these metals form compounds with boron, the work function decreases because of the possible transfer of the d electrons from the metal to the boron atoms (this stabilizes the boron configurations). Therefore, the work function of LaB_6 is lower than that of lanthanum, the work function of GdB_6 is lower than that of gadolinium, and the work function of LuB_6 is lower than that of lutetium.

The work function of terbium is 3.09 eV, whereas the work function of the hexaboride (TbB_6) has two values: 2.99 and 3.26 eV. This scatter of the work function of TbB_6 makes it impossible to determine finally the configuration of the valence electrons in terbium atoms ($4f^85d^16s^2$ or $4f^96s^2$). If $\varphi = 2.99$ eV is the correct value (this value was obtained for well-indexed samples), it follows that preference should be given to the $4f^85d^16s^2$ configuration and the similarity of the work functions of the boride and the pure metal suggests that both cited electron configurations exist in terbium hexaboride.

An allowance for the possibility of the formation of stable electron configurations in crystals makes it possible to explain in a natural manner some other features of thermionic emission which are associated directly with the processes occurring in the surface layer. This applies particularly to film systems, in which the dependence of the emission properties of the films on the electron structure of the substrate and the film itself is manifested particularly clearly. For example, when LaB_6 and GdB_6 films are deposited on tungsten, the thermionic emission at 2300-2400°K is enhanced. This effect is not observed when SmB_6 and TbB_6 are evaporated on tungsten [272].

The atoms of tungsten in the metallic phase are characterized by high values of the statistical weights of the d^5 stable configurations due to the partial transfer of the s electrons into the d band. When lanthanum is deposited on tungsten, the vacancies still available in the d band of the substrate (the statistical weight of the d^5 configurations in tungsten is about 96% and, therefore, the total weight of the vacancies is about 4%) can be filled with the $5d^1$ electrons of lanthanum. This is favored by the energy considerations because it leads to the formation of the stable d^5 states of tungsten and of the d^0 states of lanthanum, i.e., the free energy of the whole system is reduced. A considerable number of the tungsten electrons and particularly of the lanthanum electrons is basically in the free state and this is why strong emission is possible from the tungsten−lanthanum system. Similar effects are observed when gadolinium $(4f^7 5d^1 6s^2)$ is evaporated on tungsten. In this case, the transfer of the $5d^1$ electrons to tungsten favors the formation of the stable d^5 states of tungsten and of the f^7 states of gadolinium (we have said earlier that the energy stability of the half-filled f^7 states is high and that it influences strongly the thermionic emission). Similar results can be expected from the evaporation of lutetium on tungsten because stable f^{14} states can form as a result of the transfer of the $5d^1$ electrons to tungsten.

The emission from tungsten is not enhanced by the evaporation of samarium $(4f^6 5d^0 6s^2)$ because some of the 6s electrons are captured by the f shell of the samarium atoms and stable f^7 configurations are formed. Therefore, the probability of the participation of the 6s electrons in the emission decreases considerably. In the case of terbium ($4f^{9-x}5d^x6s^2$), the 6s electrons are again captured by the f shell but this results in the formation of the completely filled f^{14} configuration, which — as in the case of samarium — reduces the probability of the participation of the 6s electrons in the emission processes.

Similar conclusions can be drawn from investigations of the properties of various materials and coatings which are used as emission killers in oxide—cathode devices (these killers suppress the emission from the products of evaporation of the oxide cathode). An active layer evaporated from the oxide cathode onto the grid in a vacuum tube can be regarded as a polycrystalline semiconducting emitter consisting of barium oxide activated with free barium. It is shown in [273] that the substrate material has a strong influence on the activation process and on the highest possible activity. This influence is felt even when the layer evaporated from the cathode is about 1000 monolayers thick, i.e., when its thickness is such that the direct influence of the substrate on the emission can be excluded. Therefore, we must assume that the substrate material affects directly the adjoining layer of the emitter and that the further spread of the influence of the substrate results from the place exchange mechanism. In the final analysis, the emission suppression mechanism reduces to the capture of the emitter electrons by the substrate or the coating material. According to the available data (Table 20), the strongest emission-suppression properties are exhibited by silicon, chromium, and tungsten oxides which can reduce the grid emission much more effectively than other nonmetallic or metallike compounds.

TABLE 20. Properties of Antiemission Grid Coatings during Deposition of Evaporation Products from Oxide Cathodes

Carbide	Valence electron configuration of metal atoms	φ, eV	Oxide	Valence electron configuration of metal atoms	φ, eV
ZrC	$4d^25s^2$	1.29	Fe_2O_3	$3d^64s^2$	1.57
MoC	$4d^55s^1$	1.57	TiO_2	$3d^24s^2$	1.93
TiC	$3d^24s^2$	1.61	Cr_2O_3	$3d^54s^1$	2.28
WC	$5d^46s^2$	1.74	WO_3	$5d^46s^2$	2.43

The capture of the emitter electrons by the atoms of the emission killer occurs if such capture increases the statistical weight of the stable configurations, i.e., if it reduces the free energy of the system. The strong emission from barium is due to the dissociation of its valence s^2 electrons and, in final analysis, to the presence of mobile dissociated s electrons which are not scattered strongly on the fairly stable d states.

The strong emission from barium oxide [274] is due to the presence of electrons which migrate between the cores of the barium and oxygen atoms and are, therefore, in a weakly bound state.

When oxides of the transition metals are formed, some of the valence electrons of the metal are transferred to the oxygen atoms, which tend to acquire the stable $2s^2p^6$ configuration. This effect is responsible for the ionic component of the bonding which is very pronounced in these compounds. Since the number of the metal electrons is usually insufficient, an interaction between the oxygen atoms gives rise to the s^2p^6 configurations and to the quasistable sp^3 configurations. The electrons are exchanged continuously between these configurations and, therefore, the oxygen atoms which have, at a given moment, the sp^3 configuration may acquire the s^2p^6 states by the capture of electrons from the emitter. Thus, the antiemission properties of oxides are primarily due to the incomplete nature of their electron configurations and the low value of the statistical weight of the s^2p^6 states, i.e., due to the high statistical weights of the intermediate electron configurations. These properties are due to the fact that the transition metal cannot transfer its valence electrons to the oxygen atoms, or due to the shortage of such transferable electrons. This is why WO_3 is an excellent antiemitter: the configuration of the valence electrons in a tungsten atom ($5d^46s^2$) is responsible for the high statistical weight of the d^5 states in the compounds of tungsten and only few of the valence electrons can be transferred at random to the oxygen atoms which are characterized by low values of the statistical weights of their s^2p^6 states. Thus, the atoms of oxygen are ready to accept the emitter electrons because this reduces the free energy of the system as a whole (in such a process, the statistical weights of the stable configurations of barium or barium oxide also increase). In the case of chromium oxide (Cr_2O_3), the probability of the transfer of the 4s electrons to the oxygen atoms is somewhat higher (because of the lower principal quantum number of the d valence electrons of chromium, compared with the corresponding number of tungsten). Therefore, the antiemission properties of chromium oxide are weaker. These properties are even weaker for TiO_2, in which an atom of titanium transfers about four (s + d) electrons to the oxygen atoms and the metal tends to form the stable d^0 configurations. Very weak antiemission properties are exhibited by Fe_2O_3 because about two iron electrons are transferred to each of the oxygen atoms and the iron assumes the stable d^5 configuration. It follows that the oxygen atoms in Fe_2O_3 are unlikely to capture many emitter electrons. It is worth noting that the work function of a vacuum-tube grid coated with Fe_2O_3 (1.57 eV) differs only slightly from the work function of barium oxide emitters (1.1-1.3 eV). It is also worth noting the relatively poor antiemission properties of the transition-metal carbides. This is due to the fact that the majority of the electrons in these compounds is used up in the stabilization (increase of the statistical weight) of the sp^3 configurations of the carbon and metal atoms. Therefore, the sp^3 states in such carbides are filled and there are some excess electrons, i.e., the conditions are unfavorable for the acceptance of electrons from the emitter. This applies also to the compounds formed by the sp-type elements, such as the boron and silicon carbides, which are characterized by high values of the statistical weights of the sp^3 or the s^2p^6 configurations.

It follows from the results listed in Table 21 that the temperature of appearance of a significant thermionic emission current increases with increasing rate of capture of the emitter electrons by the grid material. The lowest temperatures at which thermionic currents appear are obtained for metals characterized by high values of the statistical weights of the d^{10} (Ni) or the d^5 (Mo, W) configurations, i.e., for those metals which cannot accept emitter electrons. Therefore, the work function is then exactly equal to the work function of the emitter evaporated on the grid. Titanium can capture few electrons (this is accompanied by an increase in the statistical weight of the d^5 configurations) and a thermionic current is emitted by this metal at a temperature which increases somewhat with the work function. Zirconium is capable of capturing more emitter electrons because of its high principal quantum number, the high energy stability of the car-

TABLE 21. Temperature of
Appearance of Significant
Thermionic Emission
Current in Vacuum Tubes
with Grids Made of Different
Materials (measurements
made during deposition
of evaporation products
from oxide cathodes)

Grid material	Temp. of appearance of thermionic current, °C	φ, eV
Ni, Mo, W	250—300	1.1—1.3
Ti, Cu, chromoberyllium bronze	300—350	1.2—1.5
Ag, Fe, Zr, Pt, Al, Cr, Pd, C	350—400	1.4—1.8
Cr_2O_3, WO_3	400—450	1.6—2.0
Au	450—500	1.8—2.0
Alloy of Au with 20% Pd	550—600	2.2—2.4

bon-type configurations formed as a result of the electron capture, and the tendency to capture emitter electrons which increase
the statistical weight of the sp^3 configurations. The ability of gold
to capture electrons is due to the high statistical weight of the d^{10}
configurations (this statistical weight is below 100%) and the tendency of this weight to increase. Silver is a much less active
antiemitter than gold, in spite of the fact that the atoms of silver
capture emitter electrons into the d states: this is because the
principal quantum number of the d states is lower and, consequently, their energy stability is less.

8. Spark Erosion Resistance

An estimate of the erosion resistance of the electrodes used
in spark machining can be obtained from the Palatnik criterion
[278]:

$$K = C\rho\lambda T_{mp}^2, \tag{32}$$

where C is the specific heat; ρ is the density; λ is the thermal
conductivity; T_{mp} is the absolute melting point of the metal in

TABLE 22. Thermal Properties, Erosion Resistances
Criterion K, and Relative Volume Wear γ
of Electrodes Used in Spark Machining

Metal	T_{mp}, °K	$\rho \times 10^{-3}$, kg/m³	C_{298}, J·kmole⁻¹·deg⁻¹	λ, W·m⁻¹·deg⁻¹	K	γ, %
W	3650	19.23	24785.86	134	$23.4 \cdot 10^{11}$	44
Cu	1356	8.92	24492.72	406	$12 \cdot 10^{10}$	190
Cr	2176	7.16	23273.61	886	$32 \cdot 10^{9}$	192
Ni	1728	8.96	26083.76	67.0	$28 \cdot 10^{9}$	466
Co	1768	8.86	24660.25	69.5	$27 \cdot 10^{9}$	292
Zn	692,7	7.14	20808.4	112	$11.6 \cdot 10^{9}$	985
Fe	1812	7.86	25078.0	73,3	$26.2 \cdot 10^{8}$	244

question. The calculated values of this criterion are listed in
Table 22.

According to Palatnik [278], the erosion resistance of a metal
improves with increasing value of K. If we analyze the results
presented in Table 22, we may assume that — in accordance with
the Palatnik criterion K — the wear of electrodes in spark machining should increase from tungsten to iron in the sequence used
in the first column of Table 22.

However, the results of more recent investigations do not
confirm this sequence. It is reported in [275, 277] that the relative volume wear of electrodes in spark machining increases from
tungsten to copper and chromium, and then to iron, cobalt, and
nickel, with the greatest wear exhibited by zinc. Obviously, the
criterion K cannot be applied to all metals and compositions but
only to some restricted group of metals.

Since the stable electron configurations are not easily disturbed by any excitation, we may assume that, other conditions
being equal, the delocalized electrons will have the highest mobility and will be excited most easily. These are the electrons
which are primarily responsible for the erosion of the electrode
in spark machining. In view of this, we may expect the erosion
resistance of the transition metals to improve with increasing
values of the statistical weight of the d^5 stable configurations, and
to deteriorate with increasing proportion of the collective-state
electrons.

It follows from the results plotted in Fig. 32 that, in the case
of spark machining in kerosene, the erosion wear decreases from

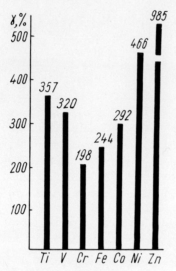

Fig. 32. Wear γ of electrodes made of d-type transition metals and used in spark machining in kerosene.

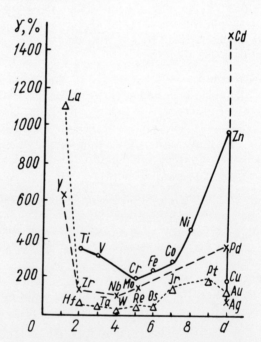

Fig. 33. Dependences of the wear of d-type transition metals in spark machining on the degree of occupation of the valence shells.

titanium to chromium and this corresponds to an increase in the statistical weight of the d^5 configurations. This is also observed in the transition from yttrium to molybdenum and from lanthanum to tungsten. The wear increases strongly in the iron-group metals if we go from iron to nickel, i.e., the wear increases with the statistical weight of the less stable d^{10} configurations. A general analysis of these results shows that the wear of various materials decreases in the transition from the first to the third long period, which corresponds to an increase in the principal quantum number of the valence electrons, i.e., to an increase in the stability of the electron configurations (Fig. 33). Thus, the highest erosion resistance among the transition metals should be exhibited by tungsten, which is indeed observed experimentally.

Experimental studies show that the lowest wear is exhibited by tungsten – copper alloys of the type used in electric contacts. The low wear of these alloys under spark machining conditions is due to the exchange of electrons between tungsten and copper, increase in the statistical weights of the d^5 and d^{10} configurations, reduction in the proportion of the delocalized electrons, and appearance of a characteristic ordering which is far less perfect than that encountered in chemical compounds or solid solutions but is sufficient to result in the deviation of the properties of the tungsten – copper alloys from the additive rule.

An examination of the erosion resistance of the electrodes made of refractory compounds of transition metals with nonmetals (the most reliable and complete data are available on carbides) shows that the resistance of the group IV metal carbides is an order of magnitude higher than the resistance of the group VI carbides. This is due to the electron interaction between the carbon and the metal. According to the lastest investigations [280], the fracture of carbide electrodes used in finishing operations is of the brittle intergranular type. Moreover, it is found that the concentration of carbon in the surface layers of the carbides decreases after prolonged use. All these observations indicate that the principal factor that determines the erosion resistance of the refractory compounds is the high statistical weight of the d^5 configurations, which is characteristic of the Me – Me bonds. This is also supported by the results plotted in Fig. 34, which shows the relative wear of some refractory carbides.

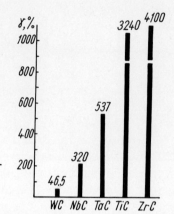

Fig. 34. Wear of refractory carbide electrodes in spark machining.

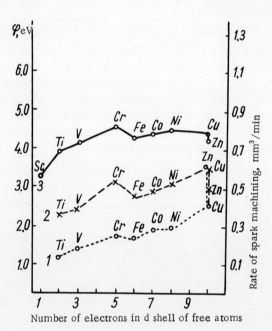

Fig. 35. Wear of the anode in spark machining, compared with the work function of the d-type transition metals: 1, 2) rate of wear measured in two different ways; 3) work function φ.

When we compare the rate of wear of an anode, plotted as a
function of the number of electrons in the d shell, with the work
function (Fig. 35) and ignore the details of the positions of the experimental points on the rising curves, we find that the rate of
wear of the anode metal increases with increasing work function
of the metal used as the cathode or, which is equivalent, with the
increasing number of electrons in the d shells and the increasing
tendency to form the stable d configurations.

These experimental results seem to be in conflict with the
earlier conclusion that the wear of the anode metal increases
with increasing flux of electrons from the cathode to the anode.
However, we must bear in mind that the rate of wear of the electrode metal in spark machining depends, other conditions being
equal, on the kinetic energy of the electrons, i.e., on their velocity from the cathode to the anode.

We may assume that the electrons traveling from the cathode
to the anode have a range of kinetic energies and that the highest
velocities are exhibited by those electrons for which the statistical weight of the atoms with localized valence electrons is highest.
These electrons have high velocities because high energies are
needed to disturb the stable configurations. This affects the emission of electrons participating in the s−d exchange process. It is
evident from Fig. 35 that the rate of wear is correlated with the
stability of the d configurations.

These ideas on the influence of the cathode material on the
rate of spark machining, which is affected by the emission of
electrons with different kinetic energies, should be considered
in conjunction with an allowance for the influence of the voltage
applied between the electrodes.

It is reported in [281] that the application of an external field
produces a channel of strongly ionized particles (a streamer) in
the gap between the electrodes. This channel grows from the
anode to the cathode or in the reverse direction. When a positive
streamer reaches the cathode and connects it to the anode, a
cathode spot is formed. A current pulse travels very rapidly
(10^9–10^{10} cm/sec) along this channel and converts it into a spark
discharge, which completes the breakdown stage of the discharge
gap. The velocity of the current pulse or the electron avalanche

along the streamer channel depends on the electric field and on
other factors. The voltage rise time increases with increasing
capacitance between the electrodes. The intensity of each spark
(pulse) and of the effects resulting from its passage also increases
with increasing capacitance.

The rate of rise and the value of the voltage across the electrodes depends also on the cathode material, i.e., on its emission
properties. The lower the work function of the electrons in the
cathode metal, the lower is the voltage that has to be applied to
the electrodes. A study of the tunnel effect [281] shows that the
probability of tunneling across a potential barrier and, consequently, the density of the cold emission current are both functions of the external field. If the work function is low, the external
field needed to produce cold emission is low.

Consequently, in the case of those metals which have low
work functions, we need a lower voltage for the extraction of an
electron from a metal into the electrode gap than the corresponding voltage needed for metals with higher work functions. It follows that the extraction of electrons from the cathode metal will
occur at different voltages for different metals and, consequently,
the electron velocities will vary with the cathode material. Thus,
the rate of wear of the anode metal depends on the number of col-
lisions of the electrons with the anode and on the momentum of
these electrons. The higher this momentum, the more rapid is the
wear in spark machining.

If the electrode used in spark machining is made of a metal
in which a considerable fraction of the valence electrons is delocalized, the rate of spark machining will be relatively high but
the wear of the electrodes will also be high. If we use metals
characterized by high values of the statistical weights of the stable configurations and a small fraction of the delocalized electrons, the wear will be less and the rate of spark machining will
increase. The rate of machining will also increase with the principal quantum number of the valence electrons of the metal atoms.

9. Superconductivity

The temperature of the transition to the superconducting state,
calculated using the Bardeen—Cooper—Schrieffer theory [282], is
related to the energy necessary for the dissociation of a Cooper

pair:

$$T_{cr} = Ce^{-\frac{1}{N(E_0)V}}, \tag{33}$$

where the coefficient C is proportional to the Debye temperature, and the power exponent is the product of the density of states on the Fermi surface and the interaction constant. If we assume that the interaction constant varies weakly with the composition, we find that the extrema in the dependence of the critical temperature on the parameters representing the electron structure (atomic number of an element, concentration of different elements in a compound, or concentration of different components in complex materials) should correspond to those compositions at which the Debye temperature or the density of states at the Fermi energy is higher. Hence, we may expect a qualitative correlation between the critical temperature and the electronic specific heat or the paramagnetic susceptibility (for paramagnetic materials). This is indeed observed in many cases. However, Eq. (33) contains an additional factor which is proportional to the Debye temperature and this complicates the variation of the critical temperature from one d-type transition metal to another. This is due to the fact that the density of states and the Debye temperature depend in different ways on the features of the electron energy spectra.

The density of states at the Fermi level, which is related to the nature of the intermediate configurations, should decrease with increasing statistical weight of the bonding d^5 configurations whereas the properties governed by the interatomic binding energy (they include the Debye temperature) should vary in the opposite manner. Therefore, in order to explain the observed dependences, it is necessary to make allowance for the influence of the density of states and of the Debye temperature, bearing in mind that the exponential dependence on the density of states is more important than the dependence on the constant C.

It follows from our discussion that good superconducting properties should be exhibited by those materials which are characterized by high statistical weights of the intermediate configurations (the d electrons make the principal contribution to the density of states at the Fermi level) and high statistical weights of the d^5 configurations. The importance of the stable sp^3 and d^5 con-

TABLE 23. Superconductivity
of ds and fds Elements

Element	Valence electron configuration	T_{cr}, °K	Type of lattice
Sc	$3\,d^1\ 4\,s^2$	Not observed	hcp
Ti	$3\,d^2\ 4\,s^2$	0.39	hcp
V	$3\,d^3\ 4\,s^2$	5.03	bcc
Cr	$3\,d^5\ 4\,s^1$	Not observed	»
Mn	$3\,d^5\ 4\,s^2$	» »	Cubic
Fe	$3\,d^6\ 4\,s^2$	» »	bcc
Co	$3\,d^7\ 4\,s^2$	» »	hcp
Ni	$3\,d^8\ 4\,s^2$	» »	fcc
Zn	$3\,d^{10}\ 4\,s^2$	0.82	hcp
Y	$4\,d^1\ 5\,s^2$	Not observed	»
Zr	$4\,d^2\ 5\,s^2$	0.55	»
Nb	$4\,d^4\ 5\,s^1$	9.17	bcc
Mo	$4\,d^5\ 5\,s^1$	0.92	»
Tc	$4\,d^6\ 5\,s^2$	7.8	hcp
Rh	$4\,d^8\ 5\,s^1$	Not observed	fcc
Pd	$4\,d^{10}\ 5\,s^0$	» »	»
Ag	$4\,d^{10}\ 5\,s^1$	» »	»
Cd	$4\,d^{10}\ 5\,s^2$	0 54	hcp
La	$5\,d^1\ 6\,s^2$	4.68—5.9	»
Hf	$5\,d^2\ 6\,s^2$	0.35	»
Ta	$5\,d^3\ 6\,s^2$	4.4	bcc
W	$5\,d^4\ 6\,s^2$	Not observed	»
Re	$5\,d^5\ 6\,s^2$	2.42	hcp
Os	$5\,d^6\ 6\,s^2$	0.65	»
Ir	$5\,d^7\ 6\,s^2$	0.14	fcc
Pt	$5\,d^9\ 6\,s^1$	Not observed	bcc
Ru	$4\,d^7\ 5\,s^1$	0.49	hcp

figurations under the conditions necessary for the transition to
the superconducting state follows also from the observation that
no superconductors have been found among the simple substances
belonging to the s-type elements, with the exception of beryllium,
which can be regarded as an s-type element only in the formal
sense. Table 23 lists the data on the superconductivity of the d-
type elements [283].

The elements of the chromium group are characterized by
high values (70-95%) of the statistical weight of the d^5 configurations and low values of the collective-state electron density (the
low statistical weight of the intermediate configurations is indicated by the low values of the electronic specific heat and the
paramagnetic susceptibility of these elements). A high proportion of the collective-state d electrons is found in metals of the
VA subgroup, which are also characterized by the highest (after
Cr, Mo, and W) statistical weight of the d^5 configurations. Thus,

the optimal conditions for the transition to the superconducting
state exist in this subgroup. This is in agreement with the experimental observations: the metals belonging to this subgroup
have the highest values of the critical temperature.

It should be stressed that all the superconducting transition
metals (with the exception of ruthenium) do not have electrons
localized in the d^{10} states, whereas the elements with relatively
high values of the statistical weight of the d^{10} configurations are
not usually superconducting. The localization of electrons in the
d^5 and sp^3 states is a characteristic of covalent bonding, i.e., such
bonding is important in superconducting substances.

The last point is supported also by the fact that superconducting
elements usually have simple crystal structures. It is evident
from Table 23 that all the metals with high critical temperatures
have high-temperature bcc modifications, and we know that these
modifications are associated with high values of the statistical
weight of the d^5 configurations and with weak antibonding influence of the delocalized electrons. Thus, the superconducting
properties of the transition metals are due to the strong localization of electrons in the most stable d^5 states and the optimal density of the d-type conduction electrons.

This also applies to the compounds formed from the transition elements. It follows that the compounds formed from the
metals characterized by low values of the statistical weights of
the stable configurations or by low Debye temperatures should
have low critical temperatures. This is indeed observed for compounds of titanium and other metals whose free atoms have few
electrons in the d shell (this does not apply to the special cases
when the formation of a compound results in an increase of the
proportion of the mobile d^n configurations and of the statistical
weight of the d^5 or sp^3 configurations).

A large group of superconducting compounds is formed by
niobium. When we go over from Nb_3Ag to Nb_3Au, we find that the
critical temperature rises from 8.2 to 11.5°K. The high critical
temperatures of these compounds can be attributed to the considerable concentration of the delocalized d-type electrons. The
increase in this temperature on going over from Nb_3Ag to Nb_3Au
is due to the higher stability of the $5d^{10}$ configurations compared
with the stability of the $4d^{10}$ states (this means that an increase in
the statistical weight of the d^5 configurations in Nb_3Au is accom-

panied by an increase in the proportion of the collective-state d-type electrons).

The critical temperature decreases along the sequence $Nb_3Al-Nb_3Ga-Nb_3In$ from 17.5 to 14.5 and 9.2°K, respectively. This is well correlated with the stability of the configurations of the second component (Al, Ga, In) which are formed as a result of one-electron transitions of the s → p type). A change in the density of the collective-state electrons has little influence on the critical temperature. As expected, the critical temperature of Nb_3Sn is higher because of the higher statistical weight of the sp^3 configurations and the fairly high density of the delocalized electrons. Unfortunately, in this series of niobium compounds we know only the critical temperature of Nb_3Ge (6.97°C), which seems to be in conflict with the variation of T_{cr} in the preceding series. This is due to the fact that the exceptionally high stability of the sp^3 configurations, which increases with decreasing principal quantum number of the sp electrons, reduces strongly the density of the conduction electrons because some of them are used in the stabilization of these configurations. This special feature of the stable configurations and the small number of the bonding electrons are the reasons why the compounds Nb_3C and Nb_3Si do not exist at all. The critical temperature of the compounds of niobium with the platinoids (Nb_3X) increases along the sequence $Nb_3Os-Nb_3Ir-Nb_3Rh-Nb_3Pt$ from 1.05 to 1.7, 2.5, and 9.2°K, respectively. In the case of osmium ($5d^66s^2$), we can expect the formation of the stable d^5 configurations and a low density of the conduction electrons, which leads to a low value of the critical temperature. In the case of iridium ($5d^76s^2$) and rhodium ($4d^85s^1$), the density of the collective-state d electrons is higher but the statistical weight of the d^{10} configurations is also higher. The high value of the statistical weight of the d^5 states and the fairly high value of the statistical weight of the intermediate configurations provide favorable conditions for the existence of superconductivity in the compounds of platinum ($5d^9s^1$) and this is why the critical temperature of Nb_3Pt is relatively high.

The critical temperatures of the compounds in the series $OsNb-IrNb-RhNb-PtNb$ are 1.4, 7.9, 4.1, and 2.4°K, respectively. The relatively low value of the critical temperature of RhNb (which is higher than that of OsNb) can be attributed to the low density of the delocalized d-type electrons although the stable d^5 configurations exist in this compound. In the case of PtNb, the electron configuration of niobium is transformed to the stable d^5 state and this alters the electron configuration of platinum in such a way that it becomes equidistant between the two stable d^5 and d^{10} states, i.e., the proportion of the intermediate d^x configurations decreases and the statistical weight of the d^5 states becomes larger.

The considerations outlined above are in full agreement with the results that follow from a comparison of the critical temperatures of the compounds of osmium, iridium, and platinum (Table 24). We can easily see that T_{cr} increases on transition from osmium compounds to iridium and platinum compounds. The relatively low values of T_{cr} observed for osmium compounds are due to the formation of the stable d^5 configurations (this circumstance alone would have increased T_{cr} but it is combined with a strong fall in the carrier density).

The accumulated extensive experimental data [284] make it possible to obtain a clearer picture of the behavior of superconductivity in refractory compounds.

TABLE 24. Critical Temperatures of Superconducting Compounds of Osmium, Iridium, Rhodium, and Platinum

Osmium		Iridium		Rhodium		Platinum	
compound	T_{cr}, °K	compound	T_{cr}, °K	compound	T_{cr}, °K	compound	T_{cr}, °K
OsTi	0.46	IrOs$_3$	0,3	Rh$_5$P$_4$	1.22	PtSi	0,93
OsNb$_3$	1.05	IrCr$_3$	0.45	RhTe$_2$	1.52	PtTa	1,0
OsNb	1.4	IrMo	1.0	RhMo	1.97	Pt$_{0.5}$W$_{0.5}$	1,45
OsZr	1.5	IrSc	1.03	RhTa	2.0	Pt$_{0.3}$Ta$_{0.7}$	1.2—1.5
Os$_3$Th$_7$	1.51	Ir$_3$Th$_7$	1.52	RhBi	2.06	Pt$_2$I	1.57
Os$_2$Nb$_3$	1.78—1.85	IrNb$_3$	1.7	Rh$_5$Ge$_3$	2.12	PtSb	2.1
OsTa	1.95	Ir$_2$Zr	4.1	Rh$_3$Th$_7$	2.15	PtNb	2.4
Os$_2$Nb	2.52	IrGe	4.7	RhNb$_3$	2.5	PtBi	1.21—2.4
Os$_2$W$_3$	2.21—3.02	IrTi$_3$	5.4	RhPb$_2$	2.66	PtPb$_4$	2.8
Os$_2$Lu	3.49	Ir$_2$Sr	5.7	RhZr	2.7	PtV$_3$	2.83
OsW	4.40	Ir$_2$Ca	4—6.15	RhBi$_3$	3.2	PtZr	3.0
OsMo	5.20	IrTh	6.5	RhW	2.64—3,37	PtNb$_3$	9.2
—	—	IrMo$_3$	6,8	RhNb	4,1	—	—
—	—	IrNb	7.9	Rh$_2$Ba	6.0	—	—
—	—	—	—	Rh$_2$Sr	6.2	—	—
—	—	—	—	Rh$_2$Ca	6,4	—	—

TABLE 25. Superconductivity of Binary Refractory Compounds

Metal	T_{cr}, °K	Silicide	T_{cr}, °K	Boride	T_{cr}, °K	Carbide	T_{cr}, °K	Nitride	T_{cr}, °K
Sc Ti	— 0.39	$ScSi_2$ Ti_5Si_3 $TiSi$ $TiSi_2$	<1.00 <1.20 <1.20 <1.20	TiB TiB_2	<1.20 <1.28	TiC	<1.20	ScN TiN	<1.40 $4.86—5.6$
Zr	0.546	Zr_4Si_3 Zr_6Si_5 $ZrSi$ Zr_4Si Sr_2Si Zr_3Si_2 $ZrSi_2$	<1.10 <1.20 <1.20 <1.20 <1.20 <1.20 <1.20	ZrB ZrB_2	$2.8—3.22$ <1.80	ZrC	<1.20	ZrN	$8.9—10.7$
Hf	0.35	$HfSi_2$	<1.20	HfB	<1.26	HfC	<1.23	HfN	6.2
V	5.03	V_3Si	17.0	VB V_2B	<1.20 <1.20	V_2C VC	<1.20 <1.20	VN	7.5
Nb	9.17	Nb_2Si Nb_5Si_3 $NbSi_2$	<1.20 <1.20 <1.20	NbB NbB_2 Nb_2B Nb_3B_4	8.25 <1.28 <1.28 <1.28	NbC Nb_2C	$6.0—11.1$ <1.98	NbN Nb_2N	15.6 <1.20
Ta	4.39	$TaSi$ Ta_5Si_2 $TaSi_2$	$4.25—4.38$ <1.20 <1.20	Ta_2B TaB Ta_3B_4 TaB_2	3.12 <1.28 <1.28 <1.20	Ta_2C TaC	<1.98 $3.3—9.7$	Ta_2N TaN	<1.20 <1.20

		Silicide		Boride		Carbide		Nitride	
Cr	—	CrSi, CrSi₂	<1.20, <1.20	CrB₂, CrB, —	<1.20, <1.28, —	Cr₄C, Cr₇C₃, Cr₃C₂	<1.20, <1.20, <1.28	CrN	<1.28
Mo	0.92, —	Mo₃Si₂, Mo₃Si, MoSi₀.₇, MoSi₂	<1.20, 1.30, 1.34, <1.20	Mo₂B, MoB, Mo₂B₅	4.74, <1.28, 1.28	Mo₂C, MoC	2.78, 7.7 –, 9.26	Mo₂N, MoN	<5.0, <12.0
W	—	W₃Si₂	2.84	W₂B, WB, W₂B₅	3.10, <1.28, <1.28	W₂C, WC	4.74–5.2, <1.28	W₂N	1.28
Re	2.42			Re₂B	2.80				
Fe	—	FeSi	<1.28						
Co	—	CoSi₂	1.22						
Ni	—	NiSi, NiSi₂	1.90, 1.00						

TABLE 25 (Cont ' d)

Metal	T_{cr}, °K	Silicide	T_{cr}, °K	Boride	T_{cr}, °K	Carbide	T_{cr}, °K	Nitride	T_{cr}, °K
Ru	0.49			Ru_2B Ru_7B_3	1.20 2.58	RuC	2.00		
Rh	—	RhSi	< 0,30	RhB Rh_7B_3	0.30 < 1.20				
Rd	—	PdSi	0.93						
Os	0,655	OsSi	< 0.30	Os_2B	< 1.02				
Ir	0.140	IrSi	< 1.02	IrB	< 1.28				
Pt	—	PtSi	0.88	PtB	1.28				
Tb	1,368	Th_3Si_2 (α) $ThSi_2$ (β) $ThSi_2$	< 1.20 3.16 2.41	ThB ThB_2	1.77 1.20	ThC	< 1.20	Th_3N_4	1,20
U	1,8 (γ)	U_3Si USi_2	< 1.10 < 0.30			UC	< 1.20	UN	1.20

An analysis of the results given in Table 25 yields the following general conclusions:

1. In the boride—carbide—nitride series of the group IV and VI transition metals, the lowest critical temperatures are observed for carbides and the highest for borides and nitrides.

2. In the compounds of the various transition metals with the same nonmetal, the probability of observing superconducting properties increases as the configuration of a free atom approaches the stable d^5 state and there is a reduction in the localization of the metal and nonmetal electrons in the most stable sp^3 configurations.

3. The highest critical temperatures are usually observed for niobium and molybdenum compounds; much more rarely are the high critical temperatures observed for vanadium, tantalum, and tungsten compounds.

4. When the relative concentration of a nonmetal in a metallike phase is reduced, the critical temperature usually falls. This follows from a comparison of the Nb_2N, Mo_2C, Mo_2N, W_2C phases with the compounds NbN, MoC, MoN, and WC, i.e., in the case of pronounced localization of electrons in the d^5 states of the metal component.

The relationships summarized above follow from an examination of the nature of the configurations which are formed in these refractory compounds and from a study of the density of those electrons which can form Cooper pairs.

The high statistical weight of the sp^3 configurations of carbon in titanium carbides reduces the density of the d electrons participating in the resonant Pauling covalent—metallic bonds [68] and the critical temperature decreases. On the other hand, an increase in the probability of the formation of the d^5 stable configurations by the metal atoms reduces the strength of the Me—X bonds, lowers the statistical weight of the sp^3 configurations, and enhances the collective-electron effects. These changes are reflected in the following way in the critical temperature:

$$ZrC\,(<1.20°\,K) - NbC\,(6.0 - 11.1°\,K),$$
$$HfC\,(<1.23°\,K) - TaC\,(3.3 - 9.7°\,K).$$

A further fall in the probability of the formation of the stable sp^3 configurations results in an even greater rise of the critical temperature:

$$NbC\,(6.0-11.1°\,K) - MoC\,(7.7-9.26°\,K),$$
$$Nb_2C\,(<1.98°\,K) - Mo_2C\,(2.78°\,K),$$
$$Ta_2C\,(<1.98°\,K) - W_2C\,(2.72-5.28°\,K).$$

The low critical temperature of WC is due to the high stability of the $6d^5$ configurations of tungsten, which is responsible for the strong rise in the statistical weight of the d^5 states. This reduces the density of the mobile (delocalized) d-type electrons. The strong rise of the stability of the d^5 configurations of tungsten also affects other physicochemical properties of the compounds formed by this element. The increase in the stability of the Me_2C phases (compounds with higher concentrations of the metal are not encountered among the group IV carbides) is a consequence of the approach of the electron configuration of the metal to the stable bonding d^5 state and of the increase in the $Me-Me$ interaction. This is the cause of the contradictory properties of WC, such as its very low melting point and its high microhardness, which distinguish this compound from the other transition-metal carbides.

The tendencies outlined above are observed more clearly in nitrides. In these compounds, the lower value of the statistical weight of the sp^3 configurations raises the probability of the formation of the d^5 states, which is accompanied by the weakening of the $Me-X$ bonds and the strengthening of the $Me-Me$ bonds. In the case of carbides, the fall in the critical temperature of the group VI transition-metal compounds is observed only for tungsten, which is characterized by the highest stability of the d^5 configurations. In the case of nitrides, this tendency is observed for molybdenum and chromium and even for tantalum, which belongs to group V. The enhancement of these tendencies and the rise of the critical temperature from carbides to nitrides follow from the nature of the electron structure of these compounds. This is observed quite clearly for almost all the refractory compounds but it is easiest to follow in the case of compounds of niobium, whose carbide is characterized by a high value of the statistical weight of the sp^3 configurations (however, this weight is lower than that for TiC). This leads to a reduction in the statistical weight of the d^5 states so that the density of the d electrons participating in the resonant metallic bonds and in conduction should be fairly high in the case of NbC. This circumstance, in combination with the fairly high strength of the Me−X bonds, is responsible for the high Debye temperature (about 700°C) and the high superconducting transition temperature of niobium carbide. When we go over from NbC to NbN, the fall in the statistical weight of the sp^3 configurations results in an increase in the density of the collective-state electrons and an enhancement of the strength of the Me−Me bonds (the strength of the Me−X bonds decreases somewhat). In the case of borides, we must take account of the fact that the values of the density of the collective-state electrons and of the binding energy are intermediate between those for MeC and MeN (this is due to the additional localization of electrons in the B−B bonds). The critical temperatures of borides should be, on the average, lower than those of nitrides but higher than those of carbides. This is confirmed by the experimental results:

NbC (6.0 - 11.1° K) — NbB (8.25° K) — NbN (15.6° K),

ZrC ($<$ 1.20° K) — ZrB (2.8 - 3.2° K) — ZrN (8.9 - 10.7° K),

HfC ($<$ 1.23° K) — HfB ($<$ 1.26° K) — HfN (6.2° K),

with the exception of possible deviations in the case of compounds of the group VI metals:

MoC (7.7 - 9 26° K) — MoB ($<$ 1.28° K) — MoN (12.0° K).

The fall of the critical temperature in the transition to the phases with higher concentrations of the metal is due to the same factors as discussed above: it is associated with the enhancement of the strength of the Me−Me bonds, which results inthe strong localization of the electrons in the covalent bonds, reduction in their density, and weakening of the strongest Me−X bonds. This mechanism is responsible for the fall of the critical temperatures in the following series of compounds:

NbC (6 0 - 11.1° K) — Nb_2C ($<$ 1,98° K),

TaC (3.3 - 3.7° K) — Ta_2C ($<$ 1.98° K),

NbN (15.6° K) — Nb_2N ($<$ 1.2° K).

Similar relationships should also be observed for the compounds of the transition metals with the nonmetals in higher per-

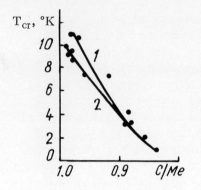

Fig. 36. Critical temperatures of niobium (1) and tantalum (2) carbides in their homogeneity regions.

iods. However, the fall in the stability of the sp^3 configurations of the nonmetal has an unfavorable influence on the superconducting properties. Therefore, compounds of this type are not usually good superconductors (Table 25).

An investigation of the dependence of the critical temperature T_{cr} on the concentration of carbon in the homogeneity regions of the carbides TaC and NbC [243] shows that the value of T_{cr} falls monotonically with decreasing carbon content (Fig. 36). The superconductivity disappears practically completely at low values of the ratio C/Me. A similar study of zirconium nitride within its homogeneity region [285] shows that when the concentration of nitrogen is reduced, the critical temperature decreases (Fig.

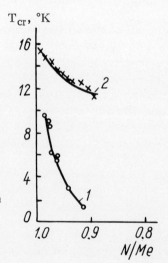

Fig. 37. Critical temperatures of zirconium (1) and niobium (2) nitrides in their homogeneity regions.

37), whereas the forbidden band width increases. The electrical
resistivity of TaC and NbC increases strongly and the temperature coefficient of the resistivity decreases when the concentration of carbon is reduced in the initial parts of the homogeneity
regions of these carbides [233]. This is due to the higher total
statistical weight of the stable configurations, which results in
stronger localization of the metal electrons in the sp^3 and d^5 configurations (Me−X and Me−Me bonds) at carbon concentrations
lower than that corresponding to the homogeneity limit. In these
circumstances, further increase in the ratio C/Me raises the
proportion of the delocalized electrons because of removal of
these electrons from the Me−Me bonds. The same effect has
been observed in the homogeneity regions of titanium and zirconium nitrides [286], in which the exchange interaction between
the metal atoms increases with decreasing concentration of nitrogen and the energy gaps which appear in the spectrum give rise
to semiconducting properties. At the same time, the forbidden
band width increases strongly and monotonically with decreasing
concentration of nitrogen. Thus, it follows from [243] that the fall
of the critical temperature with decreasing concentration of carbon in the homogeneity regions of TaC and NbC may be due to the
enhancement of the strength of the Me−Me bonds, increase in the
statistical weight of the d^5 configurations of the metal component,
and reduction in the statistical weight of the intermediate configurations.

It is interesting to consider also the data on the superconductivity of solid solutions of various compounds in one another.
The dependences of the critical temperature on the relative concentration of various carbides in MoC are plotted in Fig. 38. In
the NbC−MoC system, the highest critical temperature is observed for the solution with 50% NbC. In this solution, the stability of the d^5 configurations is so high that niobium atoms cannot
participate effectively in the formation of the covalent Me−C
bonds (this reduces the density of the collective-state electrons)
but it is insufficient for the formation of the strong Me−Me bonds
which localize a large fraction of the electrons present in pure
MoC. When we go over to the ZrC−MoC system, we find that the
critical temperature decreases in a regular manner with increasing strength of the Me−X bonds because of a regular fall in
the statistical weight of the d^5 configurations and an increase in

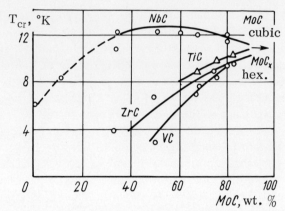

Fig. 38. Dependences of the critical temperatures of
the MoC—MeC systems on the concentration of MoC.

the statistical weight of the sp³ states. A maximum similar to that
observed for the NbC—MoC system is found also in the HfC—MoC
and TaC—MoC solutions. This maximum is related, as in the
case of the NbC—MoC system, to the reduction in the proportion
of the collective-state electrons which is observed on approach
to the compound MoC. A similar maximum is observed in the dependence of T_{cr} on the composition of the TaC—WC solid solutions (Fig. 39) but the reduction in the critical temperature at
the tungsten carbide end is stronger than in the case of niobium
carbide. This is due to the higher stability of the 6d⁵ states compared with the 5d⁵ configurations. The TaC—WC system is interesting because neither of the components is superconducting.

If we apply the configurational model, we find that there should
be three minima in the dependence of the critical temperature on

Number of valence electrons per atom

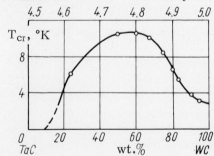

Fig. 39. Dependence of the critical
temperature of the TaC—WC system
on the concentration of WC.

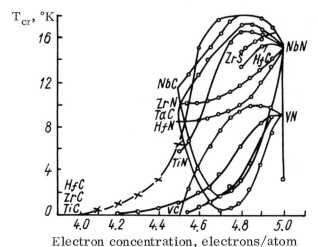

Electron concentration, electrons/atom

Fig. 40. Dependences of the critical temperature of some systems on the electron concentration (per atom).

the electron density. Two of these minima are due to very strong localization of the electrons corresponding to the highest values of the statistical weights of the sp^3 and d^5 configurations, respectively. The third minimum is of the same origin as the minimum found for the density of states in pure metals (this minimum corresponds to titanium): it is due to a reduction in the proportion of the delocalized electrons because of the localization in the stable d^5 and sp^3 configurations. These three minima should be separated by two maxima, which are usually attributed to maxima of the density of states at the Fermi level. Dependences of the critical superconducting temperature on the electron density of many systems are of this type (Fig. 40). This also applies to the dependences of the critical temperature on the electron density reported in [251], which are characterized by three minima corresponding approximately to TiC, $(Ti_{0.75}W_{0.25})C$, and $(V_{0.3}W_{0.7})C$ in the case of carbides of the group IV-VI transition metals. Further increase in the WC content in the VC−WC, NbC−WC, and TaC−WC systems raises the critical temperature, which indicates that the density of the delocalized electrons increases in spite of the high statistical weight of the stable bonding configurations.

It is clear from our discussion that an interpretation of the superconductivity on the basis of the rigid-band model, which ig-

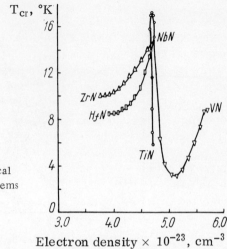

Fig. 41. Dependences of the critical
temperature of the NbN—MeN systems
on the electron density.

nores changes occurring in the electron structure as a result of
an increase in the statistical weight of the stable configurations,
sometimes fails to give the correct results. The low critical
temperature of pure scandium (which has a high density of states
at the Fermi level but a low value of the statistical weight of the
d^5 configurations) and the complex dependence of the critical temperature on the electron density in the VN—NbN system (Fig. 41),
whose components have the same electron densities but whose
metals have different stabilities of the d^5 configurations, indicate
that one should exercise great care in the use of the rigid-band
model in the explanation and prediction of the relative values of
the critical temperature of substances with complex heterodesmic
bonds.

The investigators of the physical properties of carbides, nitrides, and other refractory compounds of the transition metals
have frequently explained their results using the Bilz model [252]
of the distribution of the electron energy states. The success of
the Bilz model in the explanation of the properties of carbides and
nitrides of the groups IV and V transition metals is due to the fact
that the postulated type of binding is approximately correct for
the compounds in question because of the small covalent component in the Me—Me bonds in crystals of these compounds. However, the number of localized electrons increases as the number
of electrons in the d shell and the principal quantum number of

these electrons increase. The largest contribution of the covalent
bonding among the transition metals of groups IV-VI is found in
tungsten. The distribution of the electron states in the Me $-$ Me
energy band of a cubic carbide is similar to the distribution in
the d bands of metals because the orientation of the d_ε orbits and
the overlap of the d_ε wave functions are such that the Me $-$ Me
bonds are formed by the d electrons as well as by the s electrons
(the latter are responsible for some freedom of the electrons in
the Me $-$ Me energy band). Since the distribution of the valence
electron states in the transition metals depends on the occupancy
of the d energy band and on the principal quantum number of this
band, the distribution of the electron states in the Me $-$ Me band
of a carbide changes when one transition metal is replaced by another. This also affects the structure of the Me $-$ C band. When
we go over from $Me^{IV}C$ to $Me^V C$ and then to $Me^{VI}C$, we find that
not only is the d band gradually filled (the occupancy of this band
is low in group IV carbides) but the covalent component of the
Me $-$ Me bonds increases as well.

The covalent component of the Me $-$ Me energy band of the
transition-metal carbides is not simply equal to the statistical
weight of the d^5 configurations but these quantities are related
and an increase in the statistical weight of the d^5 states is accompanied by an increase in the covalent component of the bonding
in transition metals and, consequently, in the covalent contribution to the Me $-$ Me energy band of carbides. No serious difficulties are encountered in the interpretation of the physical properties of carbides, nitrides, and oxides of the group IV transition metals if the results are interpreted using the energy band
distribution proposed by Bilz for TiC and TiN, for which the statistical weights of the stable d^5 configurations are low. However,
when the same model is applied to the refractory compounds of
the group V transition metals, some refinements are needed which
alter slightly the shape of the energy band of the valence electron

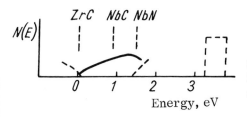

Fig. 42. Density of the electron states
in metal-like compounds (Piper
scheme).

states. In the case of the refractory compounds of the group VI d-type metals, it is necessary to introduce the idea of the formation of the covalent Me−Me bonds and the application of the rigid-band model of Bilz fails to explain the results obtained. If we compare the energy band scheme of Piper (Fig. 42) [281], which is based on the experimental data for ZrC, NbC, and NbN, with the Bilz scheme, based on theoretical LCAO calculations for TiC and TiN (Fig. 22) and confirmed by the experimental data, we can see that the maximum of the density of states in the Me−Me band predicted by the Piper scheme is located at much lower energies than the corresponding maximum in the Bilz model. In the Piper scheme, this maximum lies between 9 and 10 valence electrons per formula unit, i.e., between NbC and NbN.

According to the Bilz model, the maximum of the density of states is located at an energy corresponding to a valence electron concentration much higher than 10 electrons per formula unit. This is due to an increase in the proportion of the covalent component in the Me−Me band, which reduces the amplitude of the maximum and deforms the Me−Me band. It is obvious that, in the case of the group VI refractory compounds, the shape of the Me−Me band deviates even further from the Bilz model.

Measurements of the magnetic susceptibility of the VC−WC, TaC−WC, and NbC−Mo$_3$C$_2$ systems [251] indicate that the density of states decreases starting from 9–9.3 valence electrons per formula unit. This conclusion is based on the validity of the expression

$$\chi = 2\mu_B^2 \, N(E_0), \tag{34}$$

where μ_B is the Bohr magneton. The concentration of the conduction electrons also decreases in these solid solutions and, because the change in E_0 is slight (the number of the valence electrons per formula unit changes slightly because of an increase in the number of defects in the carbon sublattice), $N(E_0)$ decreases strongly. Therefore, we may conclude that the maximum of the Me−Me band in the energy scheme of the carbides containing the group VI metals is shifted even lower along the energy scale as a consequence of the large covalent contribution to the Me−Me band. The strong rise in the carrier density observed beginning from the composition $(Nb_{0.3}W_{0.7})C_{0.88}$ can be explained by the fact that when the valence electron concentration is 9.3 per formula unit, the d energy band intersects a higher band. In the $TaC-WC$ system, the solid solution, $(Ta_{0.3}W_{0.7})C_{0.9}$ which corresponds to approximately the same valence electron concentration, exhibits pronounced semimetallic properties, similar to those found in semiconductors (they include a negative temperature coefficient of the resistivity, a positive Hall coefficient, and diamagnetism). In the band theory, this means that the Fermi level E_0 of the solution $(Ta_{0.3}W_{0.7})C_{0.9}$ lies in the degenerate energy gap which results from the weak overlap of the tails of two neighboring bands. A comparison of the solid solutions $(Nb_{0.3}Mo_{0.7})C_{0.88}$ and $(Ta_{0.3} \cdot W_{0.7})C_{0.9}$ shows that the latter is closer to semiconductors, i.e., that it has a wider energy gap and a lower density of states in this gap. This is evidently due to the stronger localization of the electrons in the $Me-Me$ bands of tungsten and tantalum (compared with the corresponding localization in the case of niobium and molybdenum) in the formation of the covalent $Me-Me$ bonds. This is explained satisfactorily by the fact that the stability of the d^5 configurations of tantalum and tungsten are higher than the corresponding stability of niobium and molybdenum, respectively.

10. Magnetic Properties

The magnetic susceptibility of the transition metals, which are usually Pauli paramagnets, is determined by the density of states at the Fermi level. Like the electron contribution to the specific heat, the magnetic susceptibility is related to the partial transfer of the d electrons to the collective state, i.e., it is associated with a high statistical weight of excited d^n electron configurations in the intermediate spectrum.

The correlation between the magnetic susceptibility and the electronic specific heat is incomplete for the following reasons:

1. The diamagnetism of the filled shells and of the collective-state electrons makes a contribution to the magnetic susceptibility. This contribution may be considerable in the transition metals and it determines the magnetic properties of metallic copper.

2. The electronic specific heat calculated in the higher approximations includes a correction for the electron-phonon interaction, which does not occur in the expression for the paramagnetic susceptibility.

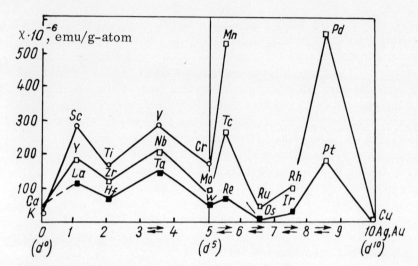

Fig. 43. Paramagnetic susceptibility of the d-type transition metals plotted against the number of electrons in the d shell.

3. A considerable contribution to the paramagnetism is made by the orbital and the Van Vleck mechanisms. These mechanisms are the result of the anisotropy of the closed shells in the d-type metals and may be comparable with the paramagnetism of the conduction electrons. A weakening of the Van Vleck paramagnetism in chromium (high values of the statistical weight of the d^5 configurations) and copper (d^{10} configurations) reduces even further the observed susceptibility of the metals. In spite of that, the qualitative correlation between C_e (Fig. 11) and χ (Fig. 43) and the correlation between the extrema of C_e and the sign of $\partial \chi / \partial T$ are evidence of the predominance of the Pauli type of paramagnetism in the metals of the long periods, which is due to the predominance of the intermediate configurations.

The nature of the strong ferromagnetism and antiferromagnetism of the transition metals is more important [145-148]. It must be stressed that, in the presence of localized moments, a magnetic order must be established in a crystal if the temperature is sufficiently low. It follows from general theoretical considerations that if there are no additional stabilization mechanisms, the antiferromagnetic order is more stable. This order is observed in the hydrogen molecule, which is evidence that out of the three possible half-filled configurations, this is the most stable arrangement. A theorem proved by Lieb and Mattis [289] shows that the ground state of a system cannot be ferromagnetic in the one-dimensional case (this also applies to some situations in the three-dimensional case). The higher stability of the antiferromagnetic state is in agreement with one of the postulates introduced by Zener [26] and is supported by the experimental information on the magnetic order. In the sole case of a magnetically ordered d-type metal (iron) and in all the magnetic transitions that occur in rare-earth metals, the high-temperature modification is always antiferromagnetic (Table 26). Zener attributes also the stability of the bcc structure to the possibility of the existence of the antiparallel arrangement of the spins of neighboring atoms.

In the band theory, a semiquantitative criterion for the existence of ferromagnetism can be given in the form

$$\Delta EN (E_0) > 1, \qquad (35)$$

TABLE 26. Magnetic Properties of Transition Metals

Element	Structure	Number of unpaired electrons	Type of unpaired electrons	Magnetic order
Cr	bcc	0.4	$3d$	Antiferromagnetic
Mn	α-phase	Depends on position	»	»
Mn	fcc	2.6	»	»
Fe	bcc	2.22	»	Ferromagnetic
Fe	fcc	0.5	»	Antiferromagnetic
Co	hcp	1.73	»	Ferromagnetic
Co	fcc	1.8	»	»
Ni	»	0.6	»	»
Ce	»	1	$4f$	Antiferromagnetic
Pr	hcp	2	»	»
Nd	»	3	»	»
Sm	»	5	»	»
Eu	bcc	7	»	»
Gd	hcp	7	»	Ferromagnetic
Tb	»	6	»	»
Dy	»	5	»	Antiferromagnetic
Ho	»	4	»	Antiferromagnetic Ferromagnetic
Er	»	3	»	Antiferromagnetic Ferromagnetic
Tu	»	2	»	Antiferromagnetic Ferromagnetic Antiferromagnetic

where $N(E_0)$ is the density of states at the Fermi level and ΔE is the exchange energy of an electron pair. This criterion is necessary but not sufficient (this applies to all the conclusions that follow from the one-electron theories). A more realistic criterion must include other characteristics of the electron structure. The density of states must be high so that the increase in the kinetic energy which results from the orientation of the electron spins is less than the reduction in the energy as a result of the exchange interaction. This is confirmed by the existence of electronic conditions in palladium which favor the transition of this metal to a strongly magnetic state because of the high density of states near the Fermi level. However, the example of palladium shows that the high density of states (which is much higher than in nickel) is of itself insufficient for the appearance of strong magnetism.

If we consider the possible mechanisms of the pairing of electron spins (Chap. I), we find that the antiferromagnetic spin arrangement is stable if a crystal has the d^0 and d^5 electron configurations, whereas the ferromagnetic order is stable if $n_d > 5$ (combinations of the d^5 and d^{10} configurations). Moreover, the existence of different types of magnetic ordering on either side of $n_d = 5$ is due to the half-filled nature of the configurations in which there are five d electrons. In the rare-earth metals, a similar situation occurs for $n_f = 7$, i.e., again we are dealing with a half-filled shell. This shows once more that the half-filled configurations are responsible for the covalent bonding which gives rise to the antiferromagnetic spin arrangement when the f^7 configurations hardly overlap. In the transition metals, the increase in the overlap from chromium to scandium increases the probability of the appearance of localized magnetic moments and the metals become Pauli paramagnets with identically filled subbands for both directions of the spin. In chromium, the magnetic moment per atom is 0.4 μ_B. This is a measure of the number of electrons which are localized at the atomic cores and were observed by Weiss and DeMarco [43]. The other electrons are distributed among the atoms and are covalently paired, i.e., they make no contribution to the magnetic moment.

The bonding mechanism, typical of the d^5 and d^{10} configurations, favors the appearance of the ferromagnetic order in the second half of the first period. The number of d electrons localized in the stable configurations and in the intermediate states in iron is 7.3, which is in agreement with the data deduced from the isomeric shift [290].*

If we ignore the intermediate spectra, we find that the statistical weights of the d^5 and d^{10} electron configurations in the iron are 54 and 46%, respectively.

The magnetic moment per one atom in iron is

$$\mu = 5P_5 \approx 2.7 \ \mu_B/\text{atom.} \qquad (36)$$

According to Vonsovskii and Vlasov [292], Mott and Stevens [25], most of the s electrons are oriented antiparallel to the spins of the 3d electrons because this reduces the energy of the s electrons as a result of the hybridization with the partly filled orbits in the d shell.† When these assumptions are made it is found that the magnetic moment is close to the experimental value of $2.2 \ \mu_B/\text{atom}$.

The advantages of this explanation of the magnetic moment of iron are as follows:

I) There is no need to postulate the existence of special electrons responsible solely for the magnetic properties.

II) The electron distribution described above also explains other properties of iron, such as the moderate value of the electronic specific heat, which is due to the fact that n_d is close to 7.5 electrons/atom; the relatively strong thermionic emission, due to the presence of the s electrons; and the crystalline allotropy.

III) The localized nature of the magnetic moment in iron, in which the intermediate configurations make a small contribution to the electron delocalization, is supported by the Langevin-type temperature dependence of the paramagnetic susceptibility of iron above the Curie point.

---

*The experimental data for iron are highly contradictory. The value used here is one of the most precise and it agrees with the proposed model. Structure investigations yield a value which is difficult to interpret from the theoretical point of view, whereas the isomeric shifts give values of n_d which are deduced from the s electron density (this makes their interpretation much easier).

†Measurements of the magnetic component of the electrical resistivity indicate the existence of a strong exchange interaction between the charge carriers and the uncompensated d electrons in iron. This is in agreement with the point of view adopted in [89].

IV) Variations of the magnetic moment with the composition of
the transition-metal alloys can be explained satisfactorily.

The statistical weight of the excited configurations in nickel
is fairly high. This is supported by the high value of the electronic specific heat, compared with that of iron. The wide intermediate spectrum and the electrons localized in the d^5 configurations give rise to a magnetic moment of 0.6 μ_B/atom in metallic
nickel. Hence, we can deduce that the localized component of the
electron concentration in nickel is n_d = 9.4 electrons/atom. The
magnetic moment is independent of the relative values of the statistical weights of the stable and intermediate configurations. In
fact, we find that

$$\mu = \Sigma (10 - i) P_i = 10 \Sigma P_i - \Sigma i P_i,$$

whereas, according to the conditions of constancy of the number
of electrons and configurations (sites) in a crystal, we have $\Sigma i P_i$ =
n_d = 9.6 and ΣP_i = 1, so that μ = 0.6 μ_B/atom, irrespective of the
values of P_i. On the other hand, the concentration of the sp electrons in nickel is low and amounts to 0.6 electron/atom. This
value is in accord with the observation that the work function of
nickel is higher than that of iron. The high statistical weight of
the d^{10} configurations hinders the hybridization of the s electrons
with d states so that these electrons occupy a s band with 0.3 elec-
tron/atom and an opposite direction of the spins. The asymmetry
of the magnetic form factor of nickel, which is due to the intermediate spectrum, is stronger than the asymmetry observed in
iron and probably stronger than that found in cobalt (a characteristic superposition of the localization and delocalization is observed in cobalt [293]. The statistical weights of the electron configurations and of the density of the collective-state sp electrons
in cobalt are in agreement with the experimental data. The electronic specific heats of iron and cobalt are practically equal,
which means that the statistical weights of the intermediate configurations in these metals are similar. The increase in the
work function and the improvement in the ability of the atoms to
accept electrons, which is observed for rising values of the statistical weight of the d^{10} configurations, indicate that the concentration of the collective-state sp electrons in cobalt is 0.65 elec-
tron/atom, which is intermediate between the concentration in
iron (0.7 electron/atom) and that found in nickel (0.6 electron/atom).

It is interesting to note that this value is almost exactly equal to the concentration of the s electrons in cobalt (0.66 electron/atom) calculated by Nathans and Paoletti [94] and by Moon [95] from the neutron diffraction data. Moreover, this value is in good agreement with the results of calculations based on isomeric shifts [290]. The increase in the delocalization of electrons from iron to nickel is also supported by the results given in [53], where it is shown that a half-filled band provides optimal conditions for the energy localization of the d electrons and that a considerable delocalization results when the band is almost filled or almost empty.

The drop in the Curie temperature from iron to nickel can also be easily explained using the configurational model. It can be attributed to the weak ferromagnetic interaction of the type described above and the low statistical weight of the "ferromagnetic" d^5 configurations. The much smaller rise in the Curie point in the transition from iron to cobalt cannot be explained without invoking additional mechanisms, although it is clear that, under the same conditions, the collective-state moments are less affected by the temperature than the localized moments because the Fermi distribution is not very sensitive to the temperature. This also explains the reduction in the number of polymorphic modifications along the sequence Mn (4) → Fe (3) → Co (2) → Ni (1). The higher number of such polymorphic modifications of Mn, Fe, and Co compared with, for example, Tc, Ru, or Rh, and the disappearance of the ferromagnetic order in metals of the second and third long periods are direct consequences of the changes in the stability of the electron configurations resulting from the increase in the principal quantum number. In the second and third long periods, the d^5 configurations can form more easily and this gives rise to two effects. The "excess" d electrons are "expelled" to the sp levels and this favors additional delocalization compared with that observed in the 3d metals. Under these conditions, the interaction between spins of different atoms can be of the antiparallel type. The first effect provides favorable conditions for the full Mott transition in a d-type metal, i.e., for the appearance of a band of paired spins and the disappearance of ferromagnetism (which is replaced by the Pauli paramagnetism). Moreover, this effect improves the conditions for the formation of hcp structures because of the establishment of the "band order." The second effect results in the narrowing of the spectrum of the intermediate configurations and a general decrease in the electronic specific heat C_e in a group of metals of the same type because the number of localized electrons approaches 5 electrons/atom.

The scarcity of experimental information on the structure of γ-Fe makes it difficult to draw definite conclusions on the nature of the magnetic moment in this modification. All that we can say is that the antiferromagnetism of γ-Fe is closely related to its hcp structure, although this structure does not provide a favorable conditions for the antiferromagnetic ordering: the statistical weight of the d^5 configurations in γ-Fe is less than in α-Fe. Consequently, the d^{10} configurations play a more important role in the formation of the crystal structure.

Little is known about the magnetic properties of the refractory compounds of the transition metals. The results of careful measurements of the magnetic susceptibility of the refractory carbides [238] are in good agreement with the predictions that follow from the configurational model. The diamagnetic susceptibility of the group IV carbides predominates over the paramagnetic component, which is in good agreement with our earlier conclusion that the statistical weight of the intermediate configurations in the $Me^{IV}C$ systems is low and that the stable states which contribute to the susceptibility predominate. For this reason, the Van Vleck paramagnetism of these compounds practically disappears.

TABLE 27. Magnetic
Susceptibility of
Refractory Carbides

Carbide	Magnetic suscepti- bility $\chi \times 10^6$ emu/g-atom	T, °K
TiC	−15.7	293
ZrC	−23.0	293
HfC	−25.3	293
VC	+26.2	193
NbC	+15.3	293
TaC	+ 9.3	293
WC	+10.0	293

When we go over from the $Me^{IV}C$ to the $Me^{VI}C$ carbides (Table 27), we find that the increase in the statistical weight of the intermediate configurations and the enhancement of the strength of the Me−Me bonds increases the paramagnetic contribution and, in accordance with [238], masks the influence of the diamagnetism of the spherically symmetric stable states and of the conduction electrons (Landau mechanism). Therefore the resultant susceptibility is positive. Moreover, the enhancement of the d localization in the metal component results in a reduction of the magnetic susceptibility of the carbides of the 4d and 5d transition metals, compared with the corresponding compounds of elements in the first long period.

In the homogeneity regions of the carbides of the group IV transition metals, the magnetic susceptibility increases with increasing number of vacancies in the carbon sublattice. The dependences $\chi = f(C/Me)$ for the $Me^{V}C$ carbides are more complex: they have a characteristic minimum at C/Me $\approx$ 0.8 (Fig. 20), which is closely related to changes in the other properties of these compounds and is explained by the special features of their configuration spectra.

The magnetic susceptibility of the complex carbides depends strongly on their composition. The solid solutions TiC−WC, ZrC−WC, and HfC−WC exhibit a susceptibility minimum at about 30 mol.% WC. The susceptibility of the complex carbides VC−WC, TaC−WC, and NbC−Mo_3C_2 decreases with increasing concentration of the group VI metal in the solid solution without any singularities. The solid solution $(Ta_{0.3}W_{0.7})C_{0.9}$ which is at the limit of the solid-solution region, has diamagnetic properties.

One of the causes of the fall in the susceptibility of the VC−WC, TaC−WC, and NbC−Mo_3C_2 solid solutions is the weakening of the paramagnetism of the cores in the metal sublattice since the magnetic susceptibility of the tungsten and molybdenum ions is less than that of the vanadium, tantalum, and niobium ions.

Moreover, the paramagnetic susceptibility of the solid solutions VC−WC, TaC−WC, and NbC−Mo_3C_2 decreases because of a fall in the density of the conduction electrons, which are used in the formation of the d^5 stable electron configurations, when the concentration of tungsten and molybdenum in the solid solution is increased (this is deduced from the Hall effect data). The diamagnetic susceptibility of the conduction electrons is only one-third of the paramagnetic susceptibility and, therefore, a change in the density of these electrons has little effect on the susceptibility.

The paramagnetic susceptibility of the complex carbides $VC-WC$, $TaC-WC$, and $NbC-Mo_3C_2$ decreases because of the increase in the covalent contribution to the bonds formed between the metal atoms. This covalent contribution is provided by electron pairs whose spins are oriented antiparallel so that the net magnetic moment of a pair vanishes. Therefore, the electron cloud of such a bonding pair should exhibit the Langevin diamagnetism, which depends on the size of the cloud, and the Van Vleck paramagnetism, which depends on the symmetry of this cloud. When the cloud becomes spherical, the paramagnetic contribution vanishes.

The magnetic susceptibility of the solid solutions $TiC-WC$, $ZrC-WC$, and $HfC-WC$ has a minimum which corresponds to about 30 mol.% WC. This minimum is associated with a minimum in the conduction electron density and a maximum in the contribution of the covalent bonding. A maximum in the composition dependence of the microhardness, which lies in the same region, can be explained easily by assuming an increase in the contribution of the covalent bonding in the solid solutions $TiC-WC$ at 20-30 mol.% WC. The dependence of the microhardness on the composition of the solid solutions $TaC-WC$ is also well correlated with the dependence of the magnetic susceptibility on the composition.

11. Mechanical Properties

The influence of the electron structure on the mechanical strength of the transition metals and their compounds and alloys can best be demonstrated by considering several specific properties.

Microhardness

According to Rebinder [294, 295] and Kuznetsov [296], the hardness of a solid is governed by its surface energy. Consequently, this property should be related to the atomic interaction

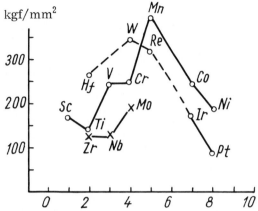

Fig. 44. Microhardness of the d-type transition metals
plotted as a function of the number of electrons in the
d shell of the free atoms.

energy which, in its turn, is a function of the electron structure
of the solid. Thus, we may expect an approximate correlation between the variation of the hardness along a series of metals and
the corresponding variations of other characteristics governed by
the atomic binding energy and the surface properties. A similar
correlation may be expected between the hardness and the statistical weight of the stable bonding configurations in a crystal.

The data plotted in Fig. 44 show that the microhardness increases along the series $Ti-V-Cr$, in accordance with the increase in the statistical weight of the stable d^5 configurations, i.e.,
those configurations which determine the covalent component in
the complex resonant covalent−metallic bonds typical of this class
of materials. This rise in the microhardness in the first half of
the first long period is repeated in other periods. The highest
value of the microhardness is observed for tungsten, in agreement with the high statistical weight and stability of the d^5 configurations in this metal. We can thus say that the microhardness
is a characteristic of a material which depends in a complex manner on other properties and is closely related to the electron
structure.

A reliable analysis of the microhardness can be made only for
crystals with the same structure (this applies also to most of the
other properties of crystals). The complex lattice of manganese,

which forms under the influence of the configuration spectrum of this metal, can influence its properties quite considerably. The high value of the microhardness of manganese is evidently due to the presence of weak as well as strong bonds between certain atoms. Since the thermal properties are sensitive to the presence of weak bonds (the intensity of the atomic vibrations may increase along such bonds), the melting point of manganese is low. The microhardness is governed by the strongest bonds, which prevent the close approach of the atoms under compression and, therefore, its value is high for manganese.

The crystals of those transition metals which have $n_d > 5$ electrons in the free state exhibit the formation of not only the d^5 but also the d^{10} configurations of the localized electrons. When n_d increases in this range, the statistical weight of the d^5 configurations decreases, whereas that of the d^{10} states increases. Bearing in mind the low stability of the d^{10} configurations, compared with the stability of the d^5 states, we may expect the bonds in the crystal lattice to become weaker when the statistical weight of the d^{10} configurations increases. This is supported by the results plotted in Fig. 44, where the transition metals with $n_d > 5$ correspond to the descending parts of the curves.

Investigations of the microhardness of the elements cannot be regarded as complete nor can the results given in Fig. 44 be regarded as completely reliable. There are as yet no data for most of the platinoids and the information on the metals in the first halves of the long periods is unreliable. The variation of the microhardness within the palladium and platinum triads should be the same as the variation of the low-temperature strength in these series. We can expect the microhardness of ruthenium and osmium to be higher than the microhardness of molybdenum and tungsten because of the large values of the statistical weights of the d^5 configurations and the additional bonding contribution of the electrons "expelled" to the sp levels. We can also expect the microhardness of molybdenum to be much lower than the values obtained for other members of this series (provided the data for chromium are reliable). This follows from the tendency for the Brinell and Vickers hardness to increase in group VI. An analysis of the available information indicates that the principal conditions for the existence of high values of the hardness of the d-type transition metals are the high statistical weight and the high stability of the bonding states of the localized electrons.

This also applies to the f-type transition metals. The generally lower values of the microhardness of these metals, compared with the d-type elements, may be explained by the fact that the f electrons do not participate directly in the formation of the bonds but the bonding is preceded by the f → d transitions. Therefore, in the condensed phase, the f-type elements behave like the d-type elements but they have lower values of the statistical weights of the d^5 stable configurations than the ordinary d-type elements. The microhardness data on the compounds plotted in Figs. 45-48

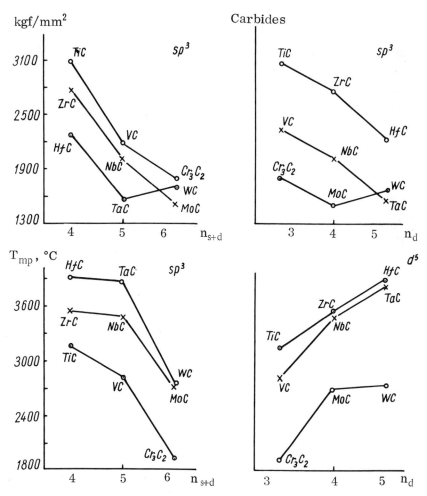

Fig. 45. Microhardness and melting points of carbides of the d-type transition metals of groups IV-VI.

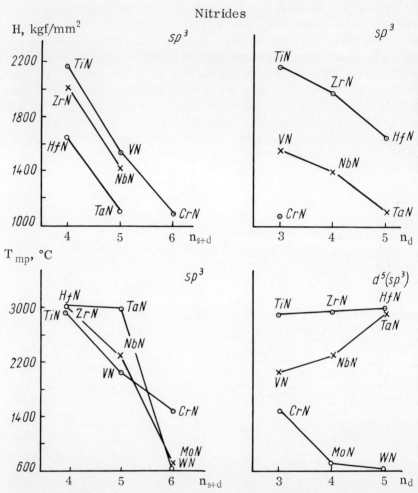

Fig. 46. Microhardness and melting points of nitrides of the d-type transition metals of groups IV-VI.

are classified in accordance with the period or group of the periodic table. The configurations which predominate in the bonding are given in each figure.

The best known compounds (carbides of the group IV-VI metals) are characterized by the highest statistical weights and stability of the sp^3 configurations formed by the carbon atoms. The Me−C interaction in these compounds results in the excitation of some of the d-type valence electrons to the sp states, which is followed by the formation of covalent bonds with the carbon atom configurations. Moreover, the donor−acceptor Me−C interaction, responsible for the partial transfer of electrons from the metal to the carbon atoms, represents the ionic component of the bonding, and the partial delocalization of the metal electrons represents the metallic component. Since the directed covalent bonds formed by the sp^3 configurations are the strongest, the binding in the carbides is determined primarily by the statistical weights of the sp^3 configurations, whereas the stability and statistical weight of the d^5 configurations, representing the covalent contribution to the Me−Me bonds, are of secondary importance. The highest value

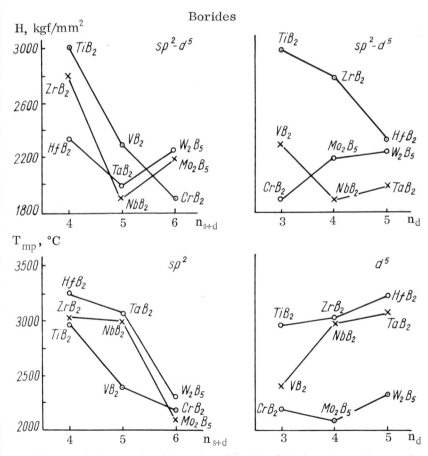

Fig. 47. Microhardness and melting points of borides of the d-type transition metals of groups IV-VI.

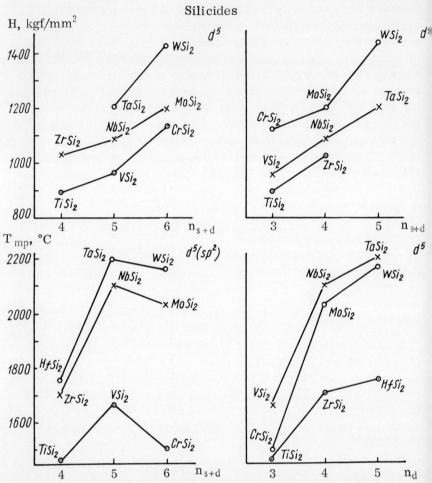

Fig. 48. Microhardness and melting points of silicides of the d-type transition metals of groups IV-VI.

of the microhardness in the carbide group is observed for the titanium compounds because the valence electrons of titanium are bound weakly in the d shell and can be excited so as to enable them to participate in the covalent $Me-C$ bonds and in the collective state (the microhardness of titanium carbides is, in fact, the highest encountered among any compounds). The statistical weight of the d^5 metallic component increases along the sequence $Ti \rightarrow V \rightarrow Cr$ and along analogs of this sequence in other groups. The sta-

tistical weight of the sp^3 configurations of the carbon atoms decreases along the same direction because of the weakening of the
stabilization of these configurations by the metal electrons and the
simultaneous reduction in the covalent component of the $Me-C$
bonds. The fall of the statistical weight of the sp^3 configurations
reduces the microhardness along the series of the carbide-forming metals (Fig. 45). A similar fall in the groups of the metal components is also due to the reduction in the statistical weight of the
sp^3 configurations (this reduction is due to an increase in the stability of the d^5 configurations). This process makes it more difficult to excite electrons so that they can participate in the $Me-C$
bonds and weakens the stabilization of the sp^3 configurations by the
delocalized metal electrons. In the third long period, some increase in the microhardness is observed only for WC, as compared with TaC (in the same period) and with MoC (in the same
group). The statistical weight of the sp^3 configurations of tungsten carbide is lower than that of its analogs but the statistical
weight and stability of the d^5 configurations is higher so that the
weakening of the $Me-C$ bonds is compensated with a margin to
spare.

When the temperature is raised, considerable changes take
place in the configuration spectrum: the d^5 configurations become
more stable, the statistical weight of the sp^3 configurations decreases, and the values of this weight become approximately equal
for different carbides. Nevertheless, the general level of the
statistical weight of the sp^3 configurations and the difference between the weights of the metal components within one period remain quite large. Therefore, the melting point (which is the high-
temperature characteristic of the mechanical strength of the lat-
tice) decreases along the series TiC $\rightarrow$ VC $\rightarrow$ Cr_3C_2 and along
their analogs when the statistical weight of the sp^3 configurations
becomes smaller (this applies also to the low-temperature microhardness). The lowest melting point is exhibited by tungsten
carbide because the melting point is governed by the strength of
the weakest bonds along which the strong lattice vibrations can be
excited.

The variation of the statistical weight of the d^5 configurations
in the groups of the carbide-forming metals is less pronounced
(even at low temperatures) than in the periods. When the temperature is raised, the differences become even smaller, and

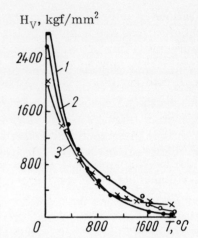

Fig. 49. Temperature dependence of the "hot" hardness of carbides of the d-type transition metals of group IV: 1) TiC; 2) ZrC; 3) HfC.

above 800°K the variation of the microhardness in a given group is governed by the increase in the stability of the d^5 configurations, which masks the influence of some fall in the statistical weight of the sp^3 states. Thus, the low-temperature variation of the hardness in the groups of the transition metals becomes reversed at high temperatures (Fig. 49). The melting points of the carbides increase in accordance with the values of the high-temperature microhardness (Fig. 45).[*]

When we go over from the carbides to the nitrides we find that the statistical weight of the sp^3 configurations of the non-metal decreases because of the corresponding increase in the statistical weight of the s^2p^6 configurations and the enlarged contribution of the ionic component to the binding in the nitrides (as compared with the carbides). The reduction in the statistical weight of the sp^3 configurations in a crystal and the weakening of the strongest covalent bonds are responsible for the general fall of the microhardness of the nitrides compared with all the carbides (this can be seen by comparing Fig. 45 with Fig. 46). The low-temperature variation of the microhardness of the nitrides within periods and groups of the transition metals is similar to that observed for the carbides, i.e., the microhardness decreases with increasing stability and statistical weight of the d^5 confi-

---

[*] The different nature of the dependences of the statistical weights of the sp^3 and d^5 configurations on the temperature can be observed even for simple elements. It results in the rise of the boiling temperature in the direction C (4470°K) → W (5645°K) although the strength of diamond is much higher than that of tungsten.

gurations of the metal (Fig. 46). At high temperatures, the sp^3 configurations are destroyed rapidly and the statistical weight of the d^5 states increases. Consequently, the strength of the Me−Me bonds, whose numerical characteristic is the statistical weight of the d^5 configurations, has a stronger influence on the melting point than on the low-temperature microhardness. An increase in the stability of these configurations within a group raises the melting point but reduces the microhardness. We must bear in mind that the nitrides of the group VI metals do not exhibit a rise but a fall of the melting point with increasing principal quantum number of the valence electrons of the metal. At first sight, this is in conflict with the increase in the statistical weight of the d^5 configurations and the enhancement of the role of the Me−Me bonds in the total binding in the lattice. In fact, this tendency is a reflection of the low values of the statistical weight of the sp^3 configurations in the nitrides of the group VI metals. The rise in the stability of the d^5 states in a group hinders the excitation of the metal electrons to the covalent Me−X bonds. This rise in the stability also reduces the ionic component of the binding and the stabilization of the sp^3 configurations. These changes in the configuration spectrum weaken very considerably the Me−X bonds in the $Me^{VI}N$ compounds and the energy of these bonds decreases along the series CrN−MoN−WN. The rise of the intensity of the vibrations of the atoms along these bonds is responsible for the fall of the thermal stability.

Further changes in the configuration spectra occur when we go over to refractory diborides. It has already been mentioned that the boron atoms in these compounds form preferentially the sp^2 configurations. The lower stability of these configurations compared with the sp^3 states of carbides and the utilization of some of the electrons in the Me−B bonds are partly compensated by the covalent interaction between the sp^2 configurations which results from the formation of the boride networks. The existence of the covalent B−B bonds (associated with the high statistical weight of the sp^2 configurations) is the reason why the microhardness of borides is higher than that of nitrides but still lower than that of carbides (Fig. 47). The Me−X bonds in borides are due to the delocalization of the metal electrons and they have the following two components: (1) the covalent component, which is due to the partial excitation of the d-type metal electrons and the formation of the hybrid spd configurations; and (2) the ionic component, which is due to the donor—acceptor interaction in the system.

The two components become weaker as we go over from $Me^{IV}B_2$ to $Me^{V}B_2$ and, therefore, the microhardness decreases along the same direction. However, further weakening of the Me—X interaction in the group VI diborides tends to increase the microhardness in the second and third long periods, whereas chromium boride has a lower microhardness than that of vanadium diboride. This variation of the microhardness is explained by the fact that the statistical weight of the d^5 configurations rises very strongly and the statistical weight of the $s^x p^y$ configurations decreases, i.e., the covalent component of the Me—B bonds becomes weaker, along the series $NbB_2 \rightarrow Mo_2B_5$ and $TaB_2 \rightarrow W_2B_5$.[*] The considerable weakening of the Me—B bonds, which gives rise to the pronounced isolation of the boron and metal sublattices, is supported by the nearly fivefold increase in the hexagonal axis in the W_2B_5 lattice compared with the TaB_2 lattice. Since the microhardness is governed by the strongest bonds in a crystal (the application of an indenter reduces the distance between the atoms in the lattice), this value is affected primarily by the increase in strength of the Me—Me bonds because of the rise in the statistical weight of the d^5 configurations and the acquisition of additional electrons by these configurations from the Me—B bonds. The melting point decreases monotonically in all borides in the transition from the $Me^{IV}B_2$ to the $Me^{VI}B_2$ ($Me_2^{VI}B_5$) compounds because of the weakening of the Me—X bonds along which strong vibrations of the lattice atoms may take place. This also applies to the microhardness values of the borides of the transitions metals which belong to the same group. In the case of the $Me^{IV}B_2$ phases, the microhardness decreases with increasing stability of the d^5 configurations, i.e., the reduction in the statistical weight of the sp^2 configurations in the system still plays the dominant role. The parrallel increase in the stability and the statistical weight of the d^5 configurations in this group is of secondary importance in the relation to the microhardness. However, in the case of the group VI carbides, the second factor is decisive and the microhardness increases with increasing principal quantum number of the d-type valence electrons along the series $CrB_2 - Mo_2B_5 - W_2B_5$.

---

[*] The replacement of Mo_2B_5 with MoB_2 and W_2B_5 with WB_2 should not alter these tendencies. The available data indicate that the properties of the MeB_2 and Me_2B_5 phases are similar.

The superposition of these two effects is observed in the group V borides: the reduction in the statistical weight of the sp^2 configurations plays the dominant role in the transition from VB_2 to NbB_2, whereas the further transition to TaB_2 corresponds to a strong rise of the stability of the d^5 configurations, which predominates over the influence of the fall of the statistical weight of the sp^2 states. Therefore, a fall in the microhardness from VB_2 to NbB_2 is replaced by a rise when we go over to TaB_2.

The further increase in the importance of the Me−Me bonds (the statistical weight of the d^5 configurations) in the transition from carbides to borides makes the variation of the microhardness within a given group approach the variation of the melting point. For example, in the case of carbides, we find an almost complete anticorrelation between the corresponding curves, whereas, in the case of borides, we find a direct correlation in the compounds of the group VI metals and partly even in the case of the Me^VB_2 phases.[*] A full correlation between the melting point and the microhardness in a given group is observed for silicides: both curves show an increase with the increasing stability of the d^5 configurations of the metal component.

The principal quantum number of the valence electrons of silicon is higher than that of carbon. This results in a strong reduction of the statistical weight of the $s^x py$ configurations of the nonmetal (silicon) and a weakening of the Me−Si bonds so that the statistical weights of the d^5 configurations and the Me−Me bonds of silicides have a decisive influence on all the strength parameters even at low temperatures (this tendency is even more marked at high temperatures). The microhardness increases along the period of the metal component of silicides (Fig. 48), which indicates that the statistical weight of the d^5 configurations increases whereas the Me−Si and Si−Si bond energies are constant or vary only slightly along the period. It must also be stressed that the fall in the melting point on transition from $Me^{IV}Si_2$ to $Me^{VI}Si_2$ and the absence of full correlation between the melting point and the microhardness in a period are due to the considerable weakening of the Me−Si bonds in chromium, molybdenum, and tungsten silicides, which intensifies the lattice vibra-

---

[*]The melting points of MoB_2 and Mo_2B_5 are lower than the melting point of CrB_2, which is in conflict with the configurational model. Therefore, further work is desirable on the refinement of the phase diagrams of the Cr−B and Mo−B systems.

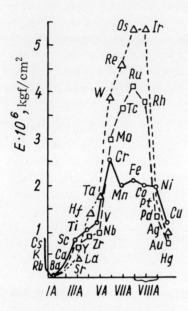

Fig. 50. Young's moduli of the d-type transition metals.

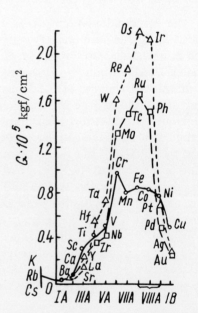

Fig. 51. Shear moduli of the d-type transition metals.

tions along these bonds. A parallel increase in the strengthening of the Me−Me bonds results in an increase in the microhardness along a period. A similar analysis can be applied to other strength parameters such as the abrasivity, mechanical strength, and brittleness. The low-temperature changes in the configuration spectra of the d-type metals are in agreement with the variation of Young's and the shear moduli (Figs. 50 and 51) which follows from the physical nature of these characteristics and from the role played by the stable configurations in determining the strength parameters of a substance. The same conclusion follows analytically from the Gilman relationship [224]

$$\sigma = E_{hkl} \frac{a}{\pi y_0} ,$$ (37)

where y_0 is the distance between consecutive atoms along a given direction and a is the radius of action of the atomic forces.

The maxima in Figs. 50 and 51 are shifted toward the group VIIIA metals because of the additional bonding contribution of the electrons "expelled" to the sp levels in the formation of the stable d^5 configurations.

The brittleness of the elements is governed by the statistical weights of the d^5, f^7, and sp^3 configurations, i.e., it increases with increasing contribution of the covalent bonding, which is related to the weight of the half-filled configurations. Moreover, the brittleness decreases with decreasing stability of these configurations and with increasing density of the collective-state electrons. In this way, brittleness gives way to plasticity, as found, for example, along the series C−Si−Ge−Sn−Pb.

The plasticity of diamond is negligibly small but it increases somewhat when we go over to silicon and germanium. By the time we reach tin, the plasticity is very high but this metal deforms with a characteristic "cry of tin," which is due to the residual sp^3 configurations of this degenerate semiconductor. When white tin is cooled to below 13°C, it transforms to the cubic diamond-like gray tin whose plasticity is low but still higher than that of silicon or germanium. Finally, we come to lead, in which the sp^3 configurations are more strongly disturbed, and we find that this is a highly plastic and very easily deformable metal. In the case of refractory compounds, the higher values of the microbrittleness should correspond to the highest values of the statistical weight of the sp^3 configurations (the most stable states) and to the strongest stabilization of these configurations by the electrons of the metal. This explains why the microbrittleness parameter[*] decreases along the series [298]:

$$TiC (9.3) — VC (6.0) — Cr_3C_2 (5.4),$$
$$TiC (9.3) — ZrC (8.0) — HfC (7.0),$$

and why similar relationships are observed for other series. However, a complete classification of the various substances in accordance with their brittleness is not yet possible because of the insufficient experimental data.

Electronic Nature of Dislocations

Dislocations are the most interesting among lattice defects because they are characteristic of a given substance and are not due to the presence of impurities. A dislocation is usually represented by the Burgers vector, which is a measure of the magnitude of energy of the distortion of the crystal lattice. Thus, the appearance, nature, and development of dislocations should be related directly to the electron structure of a substance.

Some general conclusions can be drawn from a comparison of the experimental and theoretical values of the critical shear stress $\tau_{cr} = G/2\pi$ given by I. A. Oding (Table 28). It follows from the data in this table that the formation of dislocations, which govern the ratio T/E (Table 28) is less likely in metals of the iron triad than in elements of the copper group because of the competition which occurs in the copper-group crystals between the contribution of the s configurations and the d electrons to the bonding (the statistical weight of the d^5 configurations in this metal is small). The ratio T/E increases from copper to silver because the stability of the s configurations decreases with increasing principal quantum number and the contribution of the d^5 configurations is slight because of their low statistical weight. This contribution increases and becomes predominant in gold, which has the highest principal quantum number of the d electrons. Therefore, the critical shear stress of gold is higher than the corresponding parameters of silver and copper and the ratio T/E

---

[*]The microbrittleness parameter is defined as $\gamma = (1/P_{opt})(\partial Z/\partial P)$, where P_{opt} is the optimal load and Z is the total brittleness number, introduced in [300].

TABLE 28. Comparison of
Experimental and Theoretical
Values of the Critical
Shear Stress of Some Elements

Element	Shear stress, kg/mm^2		T/E	Valence electron configuration
	experimental value, E	theoretical value, T		
Al	0.12—0.14	430	3600	$3s^2p$
Mg	0.08	280	3500	$3s^2$
β-Sn	0.13	270	2060	$5s^2p^2$
Bi	0.22	205	940	$6s^2p^3$
Cu	0.10	735	7350	$3d^{10}4s^1$
Ag	0.06	455	7600	$4d^{10}5s^1$
Au	0.09	450	5000	$5d^{10}6s^1$
Ni	0.58	1240	2130	$3d^84s^2$
Fe	2.90	1100	381	$3d^64s^2$

is smaller. It is worth noting the much lower value of the ratio T/E for iron compared with that for nickel. This is due to the increase in the statistical weight of the d^5 configurations in iron, and these configurations are more stable than the d^{10} configurations.

The dislocation density in silicon and germanium, whose atoms form energetically stable sp^3 configurations (these configurations determine the rigidity and directionality of the bonds), is 10^2-10^3 cm^{-2}, whereas the dislocation density in strongly deformed metal single crystals is 10^{11}-10^{12} cm^{-2} and in annealed metals is 10^7-10^9 cm^{-3}. Thus, the dislocation density in metals is much higher than in silicon or germanium because the stability of the d^5, d^{10}, and d^0 configurations is lower than that of the sp^3 configurations, and because the localization of the valence electrons in metals is weaker than in silicon or germanium (this is also true of the strength of the directed covalent bonds).

An analysis [30] shows that the maximum energy accumulated during plastic deformation is (in units of cal/g) 1.1 for Al, 0.5 for Cu, 1.2 for Fe, 0.8 for Ni, and 0.5 for brass. Hence we can draw the following conclusions. The greatest accumulation of the energy associated with dislocations in aluminum and iron is due to, in the first case, the low statistical weight of the sp^3 configurations and, in the second case, to the relatively high proportion of the delocalized sp electrons. The energy stored in this way decreases when we go over to nickel because of the reduction in the density of free electrons in this metal. The lowest energy is that observed for copper because of the formation of the s^2 pairs during the condensation of free atoms into a copper crystal.

Dislocations on the surface of a crystal can be revealed by etching. The points where dislocations emerge on the surface are regions where the electron configurations are strongly disturbed and this determines the physical nature of the dislocations themselves. The regions where dislocations form are governed by fluctuations so that the probability of the formation of dislocations can be reduced by careful growth or heat treatment. This approach is particularly successful when the probability of the formation of dislocations is relatively low, i.e., when the statistical weights of the more stable configurations and their energy stability are high and the number of delocalized electrons is large. Practically every crystal can be obtained in a dislocation-free state provided no fluctuations of the statistical weight of the electron configurations occur in the bulk of a crystal. This is easiest to achieve if the statistical weights are high. Since an increase in the statistical weights of the stable configurations is accompanied by an improvement in the symmetry of the crystal lattice, it follows that the dislocation density in high-symmetry crystals should be less than in low-symmetry crystals which is in agreement with the experimental observations.

Stacking Faults

There have been many attempts to explain the difference between the energies γ of the stacking faults. These attempts have been based on the crystal structure or on the assumption that the Fermi surface is in contact with the Brillouin zone boundary [302-304]. In the case of the transition metals, the energy of the stacking faults has been associated with the appearance of directed bonds [305, 306]. It is also worth considering the possibility of explaining the nature of the variations in the energy of the stacking faults in crystals on the basis of the configurational localization model [307]. Figure 52 shows the energies γ of the stacking faults in some elements (these energies are taken from the published work).

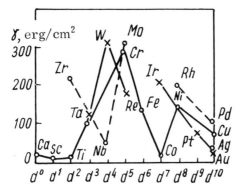

Fig. 52. Energy of stacking faults of the d-type transition metals plotted as a function of the number of d electrons per atom.

In the case of the d-type transition metals, whose atoms in the free state have $n_d \leq 5$ electrons in the d shell, we find that the crystals have the d^0 and d^5 stable configurations. The statistical weight of the d^5 configurations increases with increasing n_d along the series Sc → Ti → V → Cr. The corresponding reduction in the statistical weight of the d^0 states and of the number of electrons which go over to the delocalized state should strengthen the binding of the crystal lattice and thus increase γ. It should be pointed out that the energy of the stacking faults in niobium is low, although the statistical weight of its d^5 configurations is high (76%). The results reported in some papers [308, 309] are open to doubt; they suggest that stacking faults in niobium are due to the segregation of impurities on dislocations [310]. In crystals of those transition metals whose atoms in the free state have $n_d > 5$ electrons, the value of γ should decrease with increasing number of the valence electrons because of an increase in the statistical weight of the d^{10} configurations (the stability of the d^{10} configurations is less than that of the d^5 states).

A comparison of the stacking fault energies γ of the transition metals with the values of the low-temperature electronic specific heat shows that, in most cases, these quantities are inversely proportional [311, 312]. It has been suggested that the energy of the stacking faults in the transition metals is related to the electronic specific heat and that it is governed principally by the statistical weight of the intermediate electron configurations.

The appearance of a maximum and a minimum in the electronic specific heat in the middle of the d^0-d^5 and d^5-d^{10} intervals of the number of d electrons per metal atom is due to the low probability of the excitation of the intermediate and stable electron configurations. However, it must be pointed out that this does not apply to the 3d transition metals, such as titanium or cobalt, but it does apply to the 4d and 5d elements.

The transition metals of the groups IIIA and IVA exhibit a transition from the hcp to the bcc structure at high temperatures. This transition from a low-symmetry to a high-symmetry modification results from an increase in the statistical weight of the atoms having the d^5 configurations, which is due to the disappearance of the energetically less stable configurations. Investigations of the stacking faults in these transition metals have shown that the probability of the formation of these faults in the β modifications is lower (i.e., the energy of the stacking faults is higher) than in the α modifications.

Ytterbium ($f^{14}s^2$) is an interesting element among the lanthanoids. The energy of the stacking faults γ in this metal is approximately the same as in calcium (s^2). This is due to the fact that the f^{14} configuration in the f transition metals is very stable and, therefore, the outer s^2 electrons in ytterbium are the valence electrons. Thus, ytterbium resembles calcium in its electron structure and, therefore, the energies of the stacking faults are the same in both metals.

When alloy are formed, an exchange takes place between the stable configurations and the delocalized fraction of the valence electrons. This exchange is responsible for the binding between the stable configurations. In solid solutions, the exchange between the localized and delocalized electrons of atoms of different components "loosens" the lattice and, therefore, the energy of the stacking faults in solid solutions is less than the corresponding energy in the original components.

Howie and Swann [313] measured the energy of the stacking faults in the alloys (solid solutions) of the Cu−Zn, Cu−Al, Ag−Zn, Ag−Al, and Ni−Co systems. They found that the energy γ for the Ni−Co system fell to zero when the concentration of cobalt was increased to about 75%, i.e., when the cubic lattice transformed to the hexagonal structure. The value of γ for the copper-base alloys was higher than that for the gold-base alloys. All the alloys containing aluminum exhibited a greater fall in the energy of their stacking faults than did the alloys containing zinc. These changes in γ can be explained by the fact that the stability of the s^2 configurations decreases with increasing principal quantum number of the s-type elements (Cu, Ag, Au) and this leads to a fall in γ. In the alloys containing zinc, whose atoms have the stable $d^{10}s^2$ configurations, the probability of disturbing the stable configurations of the atoms of the other component was less than in the case of alloys with aluminum (s^2p^1). A similar relationship was observed also for the Cu−Al system. In the Ni−Co system, the statistical weight of the stable configurations fell so strongly at 75% Co that the structure changed to one with a lower symmetry. At this stage, the value of γ fell to zero.

In the Cu−Ni system, in which continuous series of solid solutions are formed. γ has a minimum for the equimolar composition [314-316], which is due to the acceptor nature of nickel, on the one hand, and the donor properties of copper, on the other.

The deformation of the Cu−Si [317], Co−Fe [318], and Cu−Ga [139] alloys gives rise to a metastable hexagonal phase and a parallel increase in the probability of the formation of stacking faults. We have already mentioned that the hcp structures are characterized by low values of the statistical weight of the stable configurations and high values of the statistical weight of the intermediate configurations. Therefore, a transition to the hexagonal structure should reduce the energy of the stacking faults γ. If we consider the hexagonal silver-base alloys, we find that the probability of the formation of stacking faults is higher in the Ag−Sn alloys than in the Ag−In alloys [320]. Since the statistical weight of the s^xp^y configurations is higher for tin than for indium, some of the s valence electrons of silver remain in the free state when the Ag−Sn alloys are formed. This makes the value of γ very small.

A study of the Cr−Fe system with bcc lattices has indicated [321] that stacking faults and coherent twins are formed in the Cr + 20% Fe alloy after filing at room temperatures, whereas no stacking faults are found in the Cr + 50% Fe alloy after a similar treatment. This is due to the fact that the statistical weight of the d^5 stable configurations of the chromium atoms reaches its highest value when 50% Fe is added (the iron atoms act as the donors of the d electrons). A study of the $Cr-Ni$ system has established [322] that a reduction in the concentration of chromium from 90 to 39% reduces the stacking fault energy γ from 81-107 to 49 ergs/cm². This is due to the fact that the reduction in the concentration of chromium increases the statistical weight of the intermediate configurations. X-ray diffraction studies of the influence of Ti, Cr, Fe, Co, Cu, and Al on the stacking fault energy in binary nickel alloys [323] has shown that all the alloying elements reduce the value of γ/G and that this reduction is 0.72 for titanium, 0.25 for chromium, 0.08 for iron, 0.14 for cobalt, 0.13 for copper, and 0.23 for aluminum. If we ignore iron, in which the formation of the d^5 and d^{10} stable configurations is unlikely, we can expect the value of γ/G to rise with increasing statistical weight of the d^5 or d^{10} stable configurations of the atoms of the alloying elements. This is not observed and, therefore, we may conclude that the changes in γ/G which occur in solid solutions are not additive.

Investigations of solid solutions with low stacking fault energies have shown that the addition of a small amount of an impurity may affect strongly the formation of stacking faults. This happens when titanium or niobium is added to austenitic chromium-nickel alloys [324]. When titanium and niobium carbides, which have the stable sp^3 configurations of the carbon atoms, are formed in these alloys, the statistical weight of the stable configurations of the matrix metals decreases [325] and, consequently, the value of γ falls. A study of the stacking faults in tantalum monocarbide in its homogeneity region ($TaCo_{0.75}$) shows that the probability of the formation of stacking faults is highest for the compositions with $C/Me < 1$ [326]. This can easily be explained by the fact that the statistical weight of the sp^3 configurations of the carbon atoms in carbides decreases with increasing concentration of bound carbon so that the $Me-Me$ bonds and the intermediate $s^x p^y$ and d^n spectra increase in importance.

Studies of the mechanism of deformation and brittle fracture of the transition-metal alloys based on the group VIA elements have shown [42] that, above the temperature T_d at which the $\sigma - T$ curves intersect, plastic slip occurs more easily than twinning; below T_d the situation is reversed. Such behavior is governed by the value of the stacking fault energy: in particular, when this energy decreases, the $\sigma - T$ curve is shifted downward and the temperature T_d increases, which confirms that the low values of the statistical weight of the stable configurations correspond to low stacking fault energies and low temperatures.

12. Recrystallization

Physically, the recrystallization process consists of overcoming the forces represented by the atomic bonds in a distorted crystal lattice, the formation of nuclei of new undistorted grains, and their subsequent growth. In this way, atoms are transferred from distorted regions to thermodynamically more stable undistorted grains [223]. Since the driving force in recrystallization is the energy accumulated during the deformation or formation of grains, all the parameters characterizing the recrystallization process should be related to the energy of the atomic interaction which itself is a function of the electron structure. Therefore, we can expect a correlation between the degree of recrystallization and other physical properties such as the electrical conductivity, the work function, or the melting point. A complete analysis of the dependence of the recrystallization parameters on the electron structure should include interpretation of the recrystallization diagrams in which the coordinates are the grain size, the degree of deformation, and the annealing temperature. However, a simple examination of the thermal characteristics of the recrystallization process shows that, like other processes, it is governed by the ability of atoms to form stable electron configurations which reduce the free energy of the system.

The deformation of a metal, which produces elastic stresses and shear within grains as well as displacements of grains with respect to one another, increases the free energy of the metal and disturbs the energetically stable states governed by the statistical weight of the stable configurations. Since a deformed metal is in a thermodynamically unstable state, it tends to return to a stable state with a lower value of the free energy. The recrystallization process includes nucleation and grain growth and is preceded by an incubation period. Initially, new grains grow at an approximately constant rate until an energetically stable state is reached as a result of the reestablishment of the electron configurations disturbed during the deformation and grain growth stages.

Let us examine the recrystallization temperatures of pure metals:

Metal	T_r, °C
Ti	500
V	800
Cr	950
Zr	550
Nb	940
Ta	1250
W	1250
Re	1500
Ru	1175
Rh	650
Ir	1200
Pt	525

The statistical weight of the stable d^5 electron configurations is lowest in titanium. Therefore, these configurations can be reestablished quite easily and therefore titanium has the lowest recrystallization temperature among the metals in the first half of the first long period. The statistical weight of the stable d^5 configurations in vanadium is higher and therefore its recrystallization temperature is higher than that of titanium (it is more difficult to reestablish a stable configuration with a higher statistical weight). The statistical weight of the d^5 configurations at temperatures corresponding to the activation of the recrystallization process, which is in the range $(0.3-0.6)T_{mp}$, is highest for chromium and, therefore, the recrystallization temperature in the Ti−V−Cr triad is highest for chromium.

When we go over to the 4d elements, we find − as expected − that the recrystallization temperatures are higher than those for elements in the preceding long period because of the increasing stability of the d^n configurations and the increasing difficulty of reestablishment of these configurations. The recrystallization temperatures of the Zr−Nb−Mo triad increase along the triad because of increasing statistical weight of the d^5 configurations. The recrystallization temperatures of the 5d metals are even higher because of the further increase in the stability of the $5d^5$ configurations compared with the $3d^5$ and $4d^5$ states.

The recrystallization temperatures of iron and nickel are 698 and 634°K respectively, i.e., the temperature T_r decreases with decreasing statistical weight of the d^5 states in these metals. It follows that the recrystallization temperature of cobalt should lie between those of iron and nickel. Under certain conditions, the tendency of a system to minimize its free energy favors additional grain growth as a result of what is known as secondary recrystallization. This is the consequence of the relatively strong disturbance of the stable configurations of the atoms at the grain boundaries and the tendency for the reestablishment of the original configurations by electron exchange between the grains. The total statistical weight of the stable configurations in a crystal increases when the area of the boundaries between grains becomes smaller. An initial excitation of the system is needed before secondary recrystallization can take place (this is also true of the primary process) and this initial excitation is frequently represented by an activation energy. The activation energy of the secondary process depends strongly on the nature of the recrystallization, the rate of heating, and the degree of plastic deformation. Thus, the experimentally determined activation energy reflects not only the nature of the substance but also the influence of various external factors. Therefore, the dependence of this energy on the statistical weight of the stable configurations may differ from the corresponding dependence of the activation energy of the diffusion process although the basic nature of the curves should be similar in both cases.

Osipov [327] put forward a hypothesis that the physical nature of the activated state is the same in the various processes that occur in solids and require thermal excitation. This is in agreement with the spirit in which all the activation processes are approached in the configurational model. Quantitatively, such an activated state can be described by the highest energy which produces "local fusion." The calculated values of such "limiting" activation energy are in satisfactory agreement with the experimental values of the activation energies of various processes in many condensed-phase materials. The limiting activation energy increases in a regular manner with increasing statistical weight of the stable d^5 configurations (Fig. 53). Moreover, the rate of increase of this energy is higher for the more stable configurations, i.e., this rate rises with the principal quantum number of the d valence electrons. Therefore, $\partial q / \partial P_5$ (q is the limiting activation energy and P_5 is the statistical weight of the d^5 states) is highest for the third long period. The metals belonging to the copper subgroup are worth mentioning specially: in the metallic state they are characterized by a high statistical weight of the d^{10} configurations so that the bonding in these metals is primarily due to the

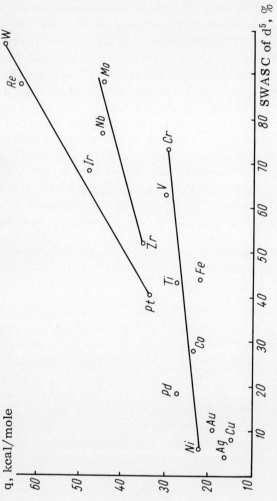

Fig. 53. Dependence of the limiting activation energy of the d-type transition metals on the statistical weight (SWASC) of the stable d^5 configurations.

s electrons which form stable s^2 pairs. Since the stability of the s configurations is much lower than the stability of the d^5 states, the limiting activation energies of these metals are very low. Since the secondary recrystallization occurs at temperatures much higher than the primary process, we find that $q_{Cr} \approx q_V$ and $q_{Mo} \approx q_{Nb}$ due to the reduction in the difference between the statistical weights of the d^5 stable configurations in the group V and VI metals at high temperatures.

Investigations of more complex systems are still in the early stages and therefore it is not yet possible to carry out a reasonably comprehensive analysis of the influence of various electronic factors which govern the recrystallization parameters. The available data on the influence of alloying on the the recrystallization temperature support the prediction that this temperature increases with the statistical weight of the d^5 states and with the degree of localization of electrons in the transition metals. Thus, alloying of molybdenum with various transition metals tends to reduce the free energy and we may expect electron transfer from the atoms of the alloying element to the d^5 configurations of molybdenum.

Such transfer is easiest to achieve in the case of zirconium, vanadium, and niobium atoms. It follows that alloying with these atoms should increase the statistical weight of the d^5 states and the recrystallization temperature much more strongly than alloying with chromium or tungsten [329]. An even stronger influence on the recrystallization temperature should result from the alloying with titanium although it is reported in [309] that titanium is less effective than zirconium. This is probably due to the use of samples of technical-grade purity [309], as indicated indirectly by the higher efficiency of the titanium admixtures (compared with that of zirconium) in the slowing down of the self-diffusion of molybdenum in titanium−molybdenum and zirconium−molybdenum alloys. In the case of niobium the recrystallization temperature is affected most strongly by the alloying with vanadium, zirconium, and titanium. This is evidently due to the strong donor tendency of these metals, which is a consequence of the low degree of localization of the valence electrons. Rhenium also gives up the delocalized electrons quite easily and these electrons are used to form stable d^5 configurations of the rhenium atoms, and to increase the statistical weight of the stable d^5 states in the matrix. Therefore, we may expect a considerable increase in the recrystallization temperature when rhenium is used as the alloying element in metals with high statistical weights of the d^5 configurations and with a strong tendency for these weights to increase. In fact, the addition of rhenium is the most effective means for slowing down recrystallization of refractory metals and alloys with bcc structures [330]. The recrystallization temperature of tungsten and molybdenum rises by about 400° when rhenium is added.

The degree of recrystallization of refractory compounds should be governed by the various contributions of the Me−Me, Me−X, and X−X bonds, which are characterized by the statistical weights of the sp^3 and d^5 stable configurations, and by the fraction of the delocalized electrons in a given compound. We must bear in mind that the sp^3 configurations are most stable from the point of view of the energy but the statistical weight of these configurations decreases strongly when the temperature is raised. Therefore, above certain temperatures the activation of recrystallization processes is basically controlled by the statistical weight of the d^5 configurations and by their stability. We have mentioned earlier that at low temperatures the microhardness, which is a characteristic of the strength of bonding in the lattice, decreases with increasing principal quantum number of the valence d electrons in the majority of those compounds in which the sp^3 configurations are fairly stable. On the other hand, the melting point rises with the principal quantum number of the metal component of a group of compounds. The parameters of the primary recrystallization should obey the high-temperature relationships, i.e., they should vary in a group in accordance with the melting point, increasing from the 3d to the 5d metals. This explains the relationship $T_r = kT_{mp}$, established by Bochvar [331] (the ratio $k = T_r/T_{mp}$ varies somewhat from one substance to another because the electron structure changes between the recrystallization temperature T_r and the melting point T_{mp}). The same approach also explains the success of Osipov's description [327] who started with the idea of "local fusion" in the various processes requiring thermal activation. The recent data on the primary recrystallization temperatures of some refractory compounds are plotted in Fig. 54 [332, 333]. The stability of the d^5 configurations and the contribution of the Me−Me bonds to the total binding energy both increase when the temperature is raised and this is why the recrystallization temperature of ZrC is higher than that of TiC. The low statistical weight of the

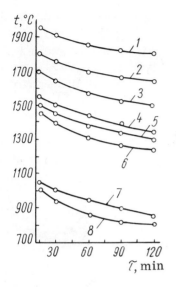

Fig. 54. Dependence of the time of onset of recrystallization in refractory compounds on the temperature at the beginning of annealing: 1) NbC; 2) ZrC; 3) TiC; 4) ZrB$_2$; 5) TiB$_2$; 6) WC; 7) Mo$_2$C; 8) MoSi$_2$.

sp^3 configurations in Mo$_2$C and WC (because of the weakening of the stability of the sp^3 configurations of carbon and the reduction in the probability of the participation of the d electrons in the co-valent Me $-$ C bonds) is reflected in the fall of the recrystalliza-tion temperatures of these two compounds as compared with the Me^{IV}C carbides. The lowest recrystallization temperature is ex-hibited by MoSi$_2$ because the stability of the sp^3 configurations of silicon is much lower than the corresponding stability of carbon. An increase in the recrystallization temperature within a group is observed also for diborides of the group IV metals whose recrystallization temperatures are generally lower than those of carbides. A comparison of the microhardness in the re-crystallized and cold-worked layers, reported in [334], demon-strates that the degree of deformation of niobium carbide is some-what higher than that of zirconium carbide, which is why the re-crystallization temperature of NbC is higher than that of ZrC. This is supported also by the values of the activation energy q of the primary recrystallization of refractory compounds, which ex-hibits the opposite tendency:

Compound	TiC	ZrC	NbC	Mo$_2$C	WC	TiB$_2$	ZrB$_2$	MoSi$_2$
q, kcal/mole	49	80	21	94	32	34	44	17

We shall conclude this section by considering the results of a recent investigation of the secondary recrystallization in some refractory carbides in their homogeneity regions [335]. The results obtained for TiC indicate a monotonic fall of the activation energy q with increasing number of carbon vacancies in the homogeneity region. We have mentioned earlier that a reduction in the concentration of carbon in titanium carbide results in a large drop in the statistical weight of the sp³ stable configurations and, consequently, in weakening of the Me − C bonds (these bonds are strongest in titanium carbide). The fall in the statistical weight of the sp³ configurations results directly from the reduction in the relative number of carbon atoms which have these sp³ configurations and from the increase in the statistical weight of the d⁵ configurations and enhancement of the strength of the Me − Me bonds, which attract additional electrons. Consequently, the stabilization of the carbon configurations by the metal electrons becomes less likely and the excitation of the d electrons to the covalent component of the Me − C bonds becomes more difficult.

13. Diffusion

The basic task of the modern theory of diffusion is to explain the observed values of the diffusion coefficient and the dependences of this coefficient on various factors, especially the temperature dependence which is described by the Arrhenius relationship

$$D = D_0 e^{-Q/kT}. \tag{38}$$

The factors that influence diffusion include the mass of the diffusing atoms, the geometrical factors (the nature of the crystal lattice), the magnetic order in a crystal, and the diffusion mechanism (vacancy or interstitial). It is obvious that the relationships governing diffusion are determined primarily by the electron structure of a substance and the factors just listed are secondary (with the exception of the isotopic mass) and it should be possible to explain them on the basis of electron structure. The low concentration of impurities in diffusion processes cannot affect greatly the properties of the matrix and, therefore, the influence of impurities on diffusion must basically be due to a change in the electron structure of the matrix. Moreover, the surfaces of solid and grain boundaries differ from the bulk crystal only because atoms with the most disturbed configurations, i.e., with low values of the statistical weight of the stable configurations, are concentrated in these regions. Hence, it follows that although the activation energy of diffusion on surfaces and grain boundaries will be less than in the bulk of the matrix, the general relationships which describe changes in diffusion from one system to another should be the same on the surface and in the bulk of the crystal. The most satisfactory theoretical approach to the quantitative description of diffusion consists of calculation of the activation energy for different mechanisms and a suitable allowance for the relative importance of these mechanisms. Such calculations are necessarily either semiempirical (in this case the results are expressed in terms of the secondary factors) or quantum-mechanical (in this case the main interactions must be allowed for because the d metals cannot be described satisfactorily by one-electron theories). It is not possible to explain the relationships governing changes in the diffusion parameters simply by considering the diffusion mechanisms because:

1) in all the mechanisms the temperature dependence of the diffusion coefficient is given by Eq. (38);
2) the parameters Q and D_0 in Eq. (38) depend on factors other than the diffusion mechanism (it is worth noting the widespread view [336] that the predominant diffusion mechanism is the same in all the transition metals or at least in all the bcc metals, although the parameters Q and D_0 vary with the atomic number);
3) the diffusion mechanism itself is governed by the electron structure of the matrix crystal.

In view of this it would be particularly interesting to determine the influence of the electron structure on the diffusion parameters of elements as well as compounds and alloys. The basic importance of such investigations was stressed by Gertsriken and Dekhtyar [337]. These authors said that "... little is known about the nature of the binding forces between atoms, and the relationship between the diffusion mechanisms and these forces has yet to be considered. This is why investigators have restricted themselves to the accumulation of data on the diffusion and self-diffusion parameters and to comparison of these data with other physicochemical constants characterizing the energy and forces of atomic binding in metals and alloys."

Diffusion is due to thermal excitation of atoms and local breaking of bonds and, therefore, it should be governed by the statistical
weight of the half-filled bonding configurations [338]. The diffusion in the transition metals is slow because of the high statistical
weight and the high stability of the d^5 configurations. The available experimental data on the diffusion in some of the transition
elements are listed in Table 29 [339]. It follows from these data
that the activation energy increases from titanium to zirconium:
this is due to the increase in the energy stability of the d configurations with increasing principal quantum number of the valence electrons. A similar increase is observed also for the group V transition metals: the activation energies Q of vanadium, niobium, and
tantalum, are 91.5, 98, and 100.4 kcal/mole, respectively. The
higher values of Q of the group V metals, compared with the group
IV metals, are due to the higher statistical weights of the d^5 stable
configurations. As expected, the group VI metals have high values
of Q, the highest activation energy being exhibited by tungsten (this
is due to the highest statistical weight and the highest stability of
the d^5 configurations in this metal). The activation energy of self-
diffusion in the metals in the first half of each of the long periods
is plotted in Fig. 55 against the statistical weight of the stable d^5
configurations. When we go over to the iron triad, we find that the
activation energy of self-diffusion decreases because of the fall in
the statistical weight of the d^5 configurations in these metals, compared with vanadium and chromium. The value of Q increases
with the principal quantum number of the valence electrons for
metals with the same type of the d^n configuration.

TABLE 29. Activation Energy of
Self-Diffusion of
Transition Metals

Element	Q, kcal/mole	Element	Q, kcal/mole	Element	Q, kcal/mole
Ti	48.0	Zr	52.0	Hf	—
V	91.5	Nb	98.0	Ta	100.4
Cr	73.2	Mo	96.9	W	120.5
Fe	64.0	Ru	—	Os	—
Co	61.9	Rh	—	Ir	—
Ni	67.0	Pd	—	Pt	66.8

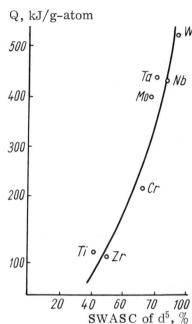

Fig. 55. Dependence of the activation energy of self-diffusion of the d-type transition metals on the statistical weight (SWASC) of the d^5 configurations.

The activation energy Q can be measured most accurately at high temperatures and, therefore, the data quoted in Table 29 refer to high-temperature diffusion. The correlation between the values of Q in Table 29 and the low-temperature statistical weights of the d^5 configurations calculated is not very satisfactory. The importance of the influence of temperature on the configuration spectra is encountered in many situations. For example, when the temperature is raised the statistical weight of the d^5 configurations increases whereas the weight of the intermediate configurations decreases. The rate of rise of the former configurations should be highest in the region of titanium for which the statistical weight of the d^5 configurations at 0°K is considerably less than the corresponding weight for vanadium. The configuration spectra change very considerably with increasing temperature and this is why the $\alpha \rightarrow \beta$ transition occurs in titanium (other properties of this metal are also affected [340]). Since the Arrhenius relationship [62] ignores changes in the electron structure of crystals [336], we find that the temperature dependence of the rate of diffusion in titanium differs strongly from the usual exponential dependence (this is often referred to as the diffusion anomaly of titanium).

The rise of the bond energies in the titanium lattice with increasing temperature is also responsible for the negative sign of the entropy of self-diffusion in this metal, i.e., for the anomalously low values of the parameter D_0. Similar factors are responsible for the diffusion "anomalies" of vanadium [341]. These anomalies are weaker in vanadium because the statistical weight of the d^5 configurations in this metal increases much less with temperature than in titanium (in particular, vanadium forms the bcc lattice at low temperatures). In the case of chromium the statistical weight of the d^5 configurations is very high even at low temperatures and, therefore, the rate of rise of this weight with temperature cannot be very rapid. The increase in the statistical weight of the d^5 states along the Ti → V → Cr series reduces the energy of the d levels in chromium so that these levels drop below the s levels. Further increase of the temperature gives rise to the d → s excitation in the system, i.e., it reduces the statistical weight of the d^5 configurations in chromium so that at sufficiently high temperatures it falls below the continuously rising statistical weight of the d^5 states in vanadium. Consequently, the activation energy of vanadium (91.5 kcal/mole) is higher than that of chromium (73.2 kcal/mole). A similar fall is also observed for those properties of chromium which are controlled by the strength of bonds at high temperatures (Fig. 9). Analogous tendencies are also observed in the second half of each of the long periods. The fall in the statistical weight of the d^{10} configurations resulting from the thermal d → s excitation of electrons results in a corresponding rise of the statistical weight of the d^5 configurations. In the case of cobalt, the statistical weight of the d^5 states decreases because of the converse s → d excitation. Thus, we find that the activation energy of self-diffusion of cobalt is 61.9 kcal/mole and that of nickel is 67 kcal/mole. Similar changes in the high-temperature relationships (compared with those applicable at low temperatures) are observed also for the various mechanical properties such as the relative values of Young's modulus [177]:

$$Fe\,(211) - Co\,(209) - Ni\,(196) \quad (T = 293°\,K);$$
$$Fe\,(167) - Co\,(188) - Ni(188) \quad T_{Fe} = 873; \quad T_{Co} = 673;$$
$$T_{Ni} = 863°\,K.$$

The rise of the stability of the d^5 configurations with increasing principal quantum number of the valence electrons and the increase in the probability of the $d \rightarrow s$ excitation in the second and third long periods weaken the effects of temperature. The dependence of the activation energy on the atomic number should be a curve with the maximum corresponding to the group VI elements with a characteristic fall in the directions of the group III and group VIIIC elements. This is indeed observed for the 5d metals for which $Q_W > Q_{Ta}$ and no significant anomalies in the dependence $Q(T)$ are found for tantalum [336]. It is also reported that the parameter D_0, which is one of the factors that determine the rate of self-diffusion, is lowest when the statistical weight of the d^5 configurations is low (Ti, Zr) or when the energy stability of these configurations is low (Cr, Zn). In all these cases the electron exchange may be quite strong but it is local because the atoms are equivalent.

It follows from our discussion that a suitable allowance for the stability, the statistical weight, and the temperature dependence of the electron configurations formed in a crystal makes it possible to understand the nature of the relationships governing self-diffusion. The activation energy is correlated with many secondary factors such as the heat of sublimation, characteristic temperature, melting point, heat of fusion, recrystallization temperature, and compressibility [342-347]. The existence of such correlations is simply due to the fact that different properties depend in similar manner on the electron spectra of crystals. The application of the configurational model shows that these correlations are of limited validity and it explains why some deviations are observed (it explains the anomalous properties of β titanium, as well as of zirconium and vanadium).

The activation energy of heterodiffusion is frequently assumed to be independent of the atomic bonding but governed primarily by the lattice structure and defects, particularly the dislocations. This view ignores the fact that the dislocation density and the energy of defect formation depend, like the lattice structure, on the electron structure of the atoms in a crystal. It is also obvious that the atomic radius or the atomic volume used in studies of various correlations also depend on the degree of localization of the valence electrons and on the energy characteristics of such localization [348].

The necessary condition for heterodiffusion is the formation of a system which is energetically more stable, i.e., which has a lower free energy, than the original system [349]. At the subatomic level this implies localization of electrons which are not localized before diffusion and appearance of additional atoms with stable electron configurations, i.e., it implies an increase in the statistical weight of the stable electron configurations. Electron exchange between atoms participating in diffusion cannot start without thermal or other excitation. This excitation can be represented by an energy E_V, which increases with increasing localization of the valence electrons and with increasing statistical weight and the stability of the preferred electron configurations. During the diffusion some of the delocalized electrons are transferred from the atoms of the diffusing substance A to the matrix atoms B. This is followed by the localization of these electrons at the cores of the B atoms and by the rise of the statistical weight of the stable electron configurations of the matrix. The result is an energy gain that can be represented by an energy E_L which is evolved during the localization and which is the thermodynamic driving force of the diffusion process. The energy balance in diffusion can be described by the difference $E_V - E_L$, which decreases if the initial localization of the valence electrons participating in diffusion becomes weaker or the final localization becomes stronger. This difference determines the activation energy of diffusion because $Q = f(E_V - E_L)$. If the localization of electrons results from the transfer of some of the delocalized electrons from an A atom to a B atom and it is described by $E_L(AB)$, it follows that $Q = f[E_V - E_L(AB)]$. The localization energy in the transfer of delocalized electrons from a B atom followed by their localization at the core of an A atom is $E_L(BA)$. If the energy associated with the AB localization is higher, this type of localization predominates and a high value of the statistical weight of the stable configurations is obtained. The BA process is also possible but it is of secondary importance because of the smaller value of the energy $E_L(BA)$. If $E_L(AB) \gg E_L(BA)$, the diffusion process is practically unipolar, as in the case of diffusion of tungsten in nickel which involves the transfer of delocalized electrons from the tungsten to the nickel atoms and an increase in the statistical weight of the d^{10} configurations of nickel while retaining the maximum value of the statistical weight of the d^5 states of tungsten.

When a diffusing element migrates in a lattice of a different substance (an atom of substance A in a B lattice), its velocity should be governed by the nature of the electron exchange between the A and B atoms. If such exchange is limited (this happens when the statistical weights of the stable electron configurations of the substances A and B — or at least one of them — are high and when the configurations formed are highly stable) the velocity of the diffusing atoms is very high. This is expressed by the preexponential factor D_0 which, like Q, depends on the electron structure and on its changes during diffusion.

This viewpoint can be used in a discussion of the parameters E and D_0 applicable to heterodiffusion in transition metals. For example, when different elements diffuse in titanium, the activation energy is governed primarily by the difficulty of the excitation of the diffusing atoms in the titanium matrix because the statistical weight of the d^5 configurations in this metal is quite high (this applies to its high-temperature β modification). Consequently, the activation energy should increase when the statistical weight of the d^5 configurations of the diffusing element becomes greater. This explains the results obtained in an experimental study of heterodiffusion in titanium [350]. The results obtained in the temperature range 900–1250°C (Table 30) indicate that the activation energy increases from titanium to vanadium and falls further along the Mn → Fe → Co series when the statistical weight of

TABLE 30. Parameters of
Diffusion of Transition
Metals in Titanium

Diffusion element	900–1250° C		1400–1650° C	
	Q, kcal/mole	D_0, cm²/sec	Q, kcal/mole	D_0, cm²/sec
Sc	32.7	$2.1 \cdot 10^{-3}$	—	—
Ti	31.2	$3.6 \cdot 10^{-4}$	45	$3 \cdot 10^{-2}$
V	32.2	$3.1 \cdot 10^{-4}$	45	$6 \cdot 10^{-2}$
Cr	35.3	$5 \cdot 10^{-3}$	42.5	$6.3 \cdot 10^{-2}$
Mn	33.7	$6.1 \cdot 10^{-3}$	39.6	$5.9 \cdot 10^{-3}$
Fe	31.6	$7.8 \cdot 10^{-3}$	38.0	$7.3 \cdot 10^{-2}$
Co	30.6	$1.2 \cdot 10^{-2}$	37.1	$1.0 \cdot 10^{-1}$
Ni	29.6	$9.2 \cdot 10^{-3}$	38.2	$2.9 \cdot 10^{-1}$

the d^5 configurations of the diffusing element increases and the coupling of this element with the matrix atoms becomes weaker.* Some fall in the activation energy of chromium, compared with that of vanadium, and its rise for nickel, compared with cobalt, can be explained by the temperature effects of the same type that are observed in self-diffusion of the same elements: at high temperatures the statistical weight of the d^5 stable configurations of vanadium is higher than that of chromium and the same is true for the nickel and cobalt pair (it should be noted that the statistical weight of the d^{10} configurations of nickel is less than that of cobalt). However, at low temperatures we find that the relationships between the activation energies are reversed ($Q_V < Q_{Cr}$ and $Q_{Ni} < Q_{Co}$), as indicated by the results given in Table 30. Electron exchange between the matrix and the diffusing element atoms decreases along the same direction (from vanadium to chromium) and, consequently, the preexponential factor rises strongly. The mobility of manganese falls considerably because of stronger electron exchange between this metal and the titanium matrix. Beyond manganese the mobility increases in the Fe−Co−Ni triad reaching its maximum value for nickel (0.29 cm^2/sec). Similar data obtained for the diffusion of various elements in niobium indicate that the influence of the term E_V on the total activation energy is greater than for the diffusion in titanium because of the higher stability of the $4d^5$ configurations (compared with the $3d^5$ states) and the higher statistical weight of the d^5 configurations of niobium compared with the corresponding weight of the same configurations in titanium. This explains the following order of the values of the activation energy Q (in kcal/mole) of heterodiffusion in niobium:

Ti (61.9); V (96.8); Fe (77.7); Co (70.5).

The activation energy for the diffusion of nickel in niobium is less than that of cobalt (65.5 kcal/mole) because it was obtained at relatively low temperatures (950-1110°C). Obviously, at higher temperatures the activation energy for the diffusion of nickel in niobium should increase faster than the corresponding energy of cobalt. On the other hand, weakening of the electron exchange resulting from the increase in the statistical weight of the d^5 con-

---

*At high pressures, which increase the statistical weight of the d^5 configurations of titanium, the activation energy of iron diffusing in titanium increases monotonically [126].

figurations is responsible for the fall of the frequency factor D_0 from vanadium to titanium and from iron to nickel (all values are given in units of cm^2/sec): Ti (5×10^{-4}); V (2.9); Fe (1.5); Co (0.74); Ni (3.5×10^{-2}). No data on the diffusion of chromium are available but a study of the diffusion of molybdenum in niobium has indicated that, in accordance with the high value of the statistical weight of the d^5 configurations and the high stability of the d^n configurations, the activation energy Q of such diffusion is the highest among the elements considered (131.2 kcal/mole) and the mobility of molybdenum atoms in the niobium lattice is high $(D_0 = 9.2 \times 10^2 \ cm^2/sec)$. The consequence of the increase in the stability of the d^n configurations in the second and third long periods is the increase of Q and D_0 within the same group of the diffusing metals: $Q_{Ti} = 61.9$ and $Q_{Zr} = 99.2$ kcal/mole; $D_{0_{Ti}} = 5.0 \times 10^{-4}$ and $D_{0_{Zr}} = 2.2 \ cm^2/sec$.

Similar relationships are observed in group V for which Q (in kcal/mole) varies in the following manner: V (96.8); Ta (99.3 ± 2.4). The published values of the activation energy for the self-diffusion of niobium are contradictory: 95.0; 96.0 $\pm$ 0.9, 105.0 $\pm$ 3, and 113.0 kcal/mole. Bearing in mind the tendencies exhibited by other metals, we can say that the correct value lies between 97 and 99 kcal/mole.

The data on the diffusion of other metals in the first halves of the long periods are incomplete but they reveal similar relationships. For example, the activation energies of tantalum, niobium, and iron are 98.7-110 [352-354], 98.7 [354], and 71.4 kcal · mole^{-1} [355], respectively. The most reliable activation energy for the diffusion of tungsten in molybdenum is 112.9 kcal/mole [356]. The scatter of the data on the self-diffusion of molybdenum (96.2-115.0 kcal/mole) suggests that the values at the upper end are somewhat overestimated.

In the second halves of the long periods the contribution of E_V decreases with decreasing statistical weight of the stable d^5 configurations and the value of E_L has a stronger influence on the activation energy. Therefore, the nature of the dependence of the activation energy on the atomic number in the transition metal series differs somewhat from the dependences discussed above. For example, when we consider the activation energies governing the diffusion of various elements in nickel, we find that the lowest value is exhibited by titanium (61.2 kcal/mole) because in this case E_V is small (the statistical weight of the d^5 configurations is low in titanium and in nickel) and the reduction in the energy as a result of localization, E_L, is large (because of the strong donor properties of titanium and the tendency of nickel to increase the statistical weight of its d^{10} configurations).

When we go over from titanium to chromium, we find that the activation energy should increase because of increase in E_V and fall in E_L with increasing statistical weight of the d^5 configurations of the diffusing element. The activation energy of manganese (66.7 kcal/mole [351]) is higher than that of titanium because E_V of manganese is larger than the corresponding parameter of titanium. No data are available for vanadium and chromium but we may assume that the activation energy of chromium is higher because E_L of chromium should be smaller than that of manganese: the high density of the delocalized electrons favors localization leading to a high value of the statistical weight of the d^5 configurations of manganese and an increase in the statistical weight of the d^{10} states of nickel. In the Fe–Co–Ni triad the activation energy does not decrease but increases monotonically because the gain E_L resulting from the localization of electrons in the stable d^{10} configurations of nickel decreases strongly and this masks the influence of the fall in the energy E_V. The highest value of the activation energy of self-diffusion among the d-type elements is exhibited by nickel for which $E_L = 0$.

At high temperatures the elements with the highest values of the statistical weights of the d^5 and d^{10} configurations may exhibit the $d \rightarrow s$ excitation of electrons and this may give rise to acceptor properties in the d shell (which is not completely filled) even in the case of elements with strong donor properties, such as silver.

The activation energies of various elements diffusing in silver are plotted in Fig. 56 as a function of the atomic number. We can see from this figure that the activation energy of heterodiffusion increases from titanium to chromium and from manganese to cobalt because the ease of transfer of the electrons from the transition metal to the d shell of silver decreases along this direction. This is observed also for the second and third long periods and for alloys of the transition metals: in these cases the activation energy decreases with decreasing statistical weight

Q, kcal/mole

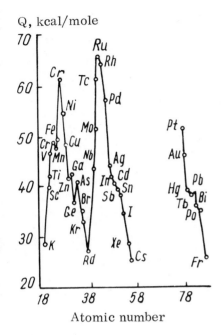

Atomic number

Fig. 56. Dependence of the activation energy Q of diffusion of various elements in silver.

and the stability of the d^5 configurations and with increasing gain in the energy of the system as a result of the increase of the total statistical weight of the stable configurations. For example, in the Nb— Zr system the activation energy of niobium varies in the following manner [358]: 108 kcal/g-atom for Nb + 5.1 at.% Zr; 89 kcal/g-atom for Nb + 15.25% Zr; 79 kcal/g-atom for Nb + 36.9% Zr. The frequency factors D_0 decrease in the same direction because of enhancement of the electron exchange which results from decrease in the statistical weight of the d^5 configurations in the alloys: the frequency factor for the three alloys just mentioned is 63, 29, and 3.3 cm²/sec. On the other hand, an increase in the concentration of molybdenum in niobium raises the activation energy and the diffusion coefficient because of the increase in E_L and the weakening of the electron exchange between the atoms of the diffusing element and the matrix. The temperature dependence of the diffusion coefficient of niobium is then given by the following equations [356]:

$$Nb : D = 9.2 \cdot 10^2 \exp(- 131.2/RT);$$
$$Nb + 16.2 \text{ at.}\% Mo : D = 1.1 \cdot 10^3 \exp(- 134.6/RT);$$
$$Nb + 19.9 \text{ at.}\% \ Mo : D = 1.4 \cdot 10^3 \ \exp(-136.5/RT).$$

TABLE 31. Activation
Energy of Diffusion in
Transition-Metal Alloys

System A–B	Concentration of B, at.%	Q(A), kcal/mole
Co—Cr	0	62.5
	5	76.0
	10	88.0
	15	95.0
	20	103.0
	25	110.0
Co—Al	0	62.5
	10	67.5
	42	85.0
	49	102.0
	50	105.0
Ni—Mo	0	64.0
	5.14	65.8
	6.9	69.0
	10.3	79.6
Ni—Cr	0	64.0
	5	61.5
	8	65.5
	10	69.0
	12	73.5
	15	80.5
Fe—Cr	0	68.0
	3.5	75.0
	7.3	90.0

    Table 31 gives the results of measurements of the activation
energy of a large batch of chromium and nickel-base alloys, carried out by Dekhtyar et al. Dekhtyar was the first to explain variations in the activation energy on the basis of electron structure: he used his own occupancy criterion [359]. This criterion
is deduced from measurements of the atomic magnetic moments
in alloys and it correlates with many physical properties as well
as with the results of neutron diffraction and x-ray spectroscopy
[357]. The arguments use by Dekhtyar are similar to those that
follow from the configurational model because the changes in the
occupancy factor in magnetic systems (provided this factor can be

determined correctly) is similar to changes in the statistical
weight of the stable configurations. This follows from the relationship between the occupancy factor q and the statistical weight
of the d^5 configurations P_5:

$$q = \frac{5P_5}{d_0}, \qquad (39)$$

where d_0 is the number of unpaired d electrons in a free atom of
a component of an alloy. In the Co–Cr system the activation energy of self-diffusion of cobalt and chromium increases with increasing concentration of chromium due to the increase in the
statistical weight of the d^5 configurations in this system, which
raises E_V and strengthens the bonds in the lattice, and due to the
fall in the energy gained as a result of the localization of the cobalt electrons in d^{10} configurations. Similar variation is observed
for the Co–Al, Ni–Al, Ni–Mo, Ni–Cr, Fe–Cr systems in which
an increase in the concentration of cobalt, molybdenum, or chromium tends to increase the statistical weight of the d^5 configurations in the crystal and the activation energy of heterodiffusion
because of higher values of E_V.

When the transition metals diffuse in the sp elements, the
value of E_L can be very high. We must bear in mind that the
transition metals are donors with respect to the sp elements and
that they can supply electrons for increase in the statistical
weight of the more stable sp^3 and s^2p^6 configurations in the system, which is accompanied by an increase E_L. Thus, when iron
diffuses in aluminum, the atoms of iron tend to transfer some of
their valence electrons to the atoms of aluminum and this results
in the formation of the energetically stable d^5 (Fe) and sp^3 (Al)
configurations. The transition to cobalt results in an increase in
the statistical weight of the d^{10} configurations, weaker electron
transfer, and fall in the value of E_L. When we go over to nickel,
we find that the value of E_L decreases strongly because of the
further stabilization of the d^{10} configurations in nickel compared
with cobalt (the available data apply to low temperatures 623-
303°K [177]). For this reason the transition from iron to nickel
results in an increase in the activation energy of self-diffusion:
this energy is 12.4, 15.6, and 19.8 kcal/mole for iron, cobalt, and
nickel, respectively. The diffusion coefficient varies in the same
way $(4.1 \times 10^{-9}, 29 \times 10^{-8}, \text{ and } 11.1 \times 10^{-6} \text{ cm}^2/\text{sec}$ for iron,

cobalt, and nickel, respectively). The increase in the diffusion coefficient is the result of enhancement of the localization and reduction in the electron exchange between the atoms. A similar increase in the activation energy and the diffusion coefficient is also observed for the case of diffusion of aluminum in the transition metals [177].

The role of the stable electron configurations in the reactive diffusion has been considered in detail in [360-362]. The kinetics of such diffusion is governed by the three main mechanisms [363], which are the adsorption, the nucleation of crystals of a chemical compound, and the growth of these crystals.

The necessary condition for each of those mechanisms and the process as a whole is the eventual formation of the system which is energetically more stable than the initial system before diffusion. The reduction in the free energy is mainly due to the interaction which occurs between the outer electron shells of the atoms of the reacting substances and which produces electron configurations more stable than the configurations of the components of the interacting system.

During adsorption the adsorbate atoms captured by the periodic field of the crystal lattice of the metal approach the adsorbent atoms so that they become subject to short-range chemical forces, i.e., they approach sufficiently closely for the overlap of the valence electron orbits. The interaction that takes place between the adsorbate and the adsorbent atoms is of the type that ensures the highest gain in energy. A process of this type results in displacements of electrons between the substrate and the adsorbed film so that fairly strong chemical bonds are established between the adsorbate atoms and the surface of the metal. The nature of the bonds is determined by the type of stable electron configurations which are formed during this process and by the density of the delocalized electrons of the reacting components.

The second mechanism – the nucleation of crystals of a chemical compound – can be divided into two stages: the reaction between the adsorbed and the matrix (substrate) atoms and the modification of the original lattice.

Since the reaction diffusion occurs at temperatures such that the mobility of atoms is high, the first stage can be described as involving transitions of atoms from the adsorbed layer to the nearest unit cells in the crystal lattice of the substrate (these atoms may occupy regular sites or interstices), followed by further penetration into the lattice (by diffusion) and arrival of new atoms from the adsorbate layer. From the thermodynamic point of view the driving force in this process is the reduction in the free energy resulting from the formation of a solid solution due to a redistribution of the electron density between the substrate and the diffusing atoms in such a way that the most stable electron configurations are formed. This produces a solid-solution zone of some finite depth and with a smooth concentration gradient.

The statistical weight of the sp^3 or the d^5 configurations increases and the energy of the system decreases with increasing amount of the diffusing element in the substrate. This increase in the concentration of the diffusing element continues until the energy liberated as a result of the increase in the statistical weight of the stable configurations becomes comparable (in local microregions) with the energy necessary for the formation of an interface between a nucleus of the new phase and the original matrix. Atoms with the most disturbed and the energetically unfavorable configurations will concentrate on this interface. The configurations that are formed within microregions and their number and distribution are of the type typical of the new phase although they are formed from the lattice of the original matrix. Therefore, the matrix lattice must become unstable: the lattice of the new chemical compound will differ from the original not only in the structure but also in the nature of chemical bonds. The moment of appearance of the new lattice represents the second stage of the formation of a nucleus of the new chemical compound.

Atoms of the diffusing component which collide with the new interface can interact (although with a lower probability than in the case of a pure metal) with the unsaturated bonds of the disturbed atomic configurations on the interface or they may interact with the collective-state electrons. This represents the third mechanism of the reactive diffusion which is the growth of crystals of the new compound. It is evident from the configurational model [364] that in reactive diffusion the energy considerations favor the migration of atoms of each kind along vacancies in their own sublattice because the formation of this sublattice results in the highest possible increase in the statistical weight of the stable configurations in a system formed from a given diffusing element and a given matrix.

Thus, the tendency for the statistical weight of the stable configurations to increase in a system determines the direction and
intensity of the reactive diffusion during all its stages. The localization of electrons in the stable states increases continuously
throughout this process and reaches its maximum value in the final
stage of the reaction. Therefore, the activation energy of the reactive diffusion can be represented in the same form as the corresponding energy of the heterodiffusion: $Q = f(E_V - E_L)$, where E_L
represents the reduction in the energy because of the stabilization
of the electron configurations as a result of formation of a new
compound. The strong dependence of the activation energy on E_L
explains the frequently observed correlation between the activation energy and the heat of formation of compounds. It has been
established [337] that the phase that appears first during reactive
diffusion is that which has the highest heat of formation although
there are some deviations from this correlation because of the
influence of E_V on the activation energy. The frequency factor is
determined, as in the case of heterodiffusion, by the rate of electron transitions between the diffusion partners.

It is important to stress the difference between the reactive
diffusion and the diffusion which does not produce a new compound.
This is easily illustrated by considering the diffusion of carbon in a
metal and in a carbide. In both cases the diffusion is accompanied by
an increase in the statistical weight of the sp^3 configurations
(which are the most stable) and the consequent redistribution of
the electron density. When carbon diffuses in a metal the increase in the statistical weight of the sp^3 configurations takes
place mainly at the expense of the delocalized electrons of the
metal (these electrons stabilize the sp^3 configurations of carbon).
The gain in the energy E_L decreases from the group IV to the
group VI transition metals with decreasing density of the delocalized electrons of the metal component (Table 32). The statistical weight of the d^5 configurations is not greatly affected.

When the diffusion process produces a compound, there are
additional ways in which the statistical weight of the sp^3 configurations can increase (these ways are related) the formation of
metal−carbon bonds typical of carbides. Apart from the stabilization of the sp^3 configurations by the collective-state electrons and
by the ionic component of the bonds (which is of the donor−accep-
tor origin), covalent bonds must be formed as a result of further

TABLE 32. Parameters
of Diffusion of Carbon
in Transition Metals

Metal	Q, kcal/mole	D_0, cm^2/sec
Ti	20.0	$3.02 \cdot 10^{-3}$
Zr	26.7	$4.8 \cdot 10^{-3}$
Hf	40.0	$4.2 \cdot 10^{-2}$
Nb	35.0	$9.3 \cdot 10^{-3}$
Ta	43.0	$2.57 \cdot 10^{-2}$

interaction and this is accompanied by an increase in the statistical
weight of the sp^3 configurations of carbon and a decrease in the
statistical weight of the d^5 configurations of the metal. The covalent
bonding produces a lattice typical of the compound and reduces
the free energy of the system. Such a diffusion process alters the
nature of the configuration spectrum and redistributes the localized
valence electrons so that a large excitation energy E_V is required.
Consequently, the activation energy Q for the reactive diffusion of
carbon in carbides is higher than the activation energy necessary
for the diffusion in metals. For the same reason (high values of
E_V) the activation energy needed for the formation of TiC by diffusion is higher than that needed to form TiB$_2$, the activation energy
for the formation of ZrC is higher than that for ZrB$_2$, and the activation energy for the formation of silicide and nitride films on
transition metals is usually much lower [238]. On the other hand,
an increase of the statistical weight of the sp^3 configurations in a
system increases the gain in the energy E_L. This gain is greatest
for the diffusion in titanium because electrons in this metal are

TABLE 33. Parameters of
Reactive Diffusion of Carbon
in Transition Metals

Phase produced	Q, kcal/mole	D_0, cm^2/sec	Temperature range, °C
TiC	62	0 1	1290—1490
ZrC	78.7	0.95	2000—2800
NbC	88.2	7.6	1700—2300
TaC	98.8	8.8	2215—2745

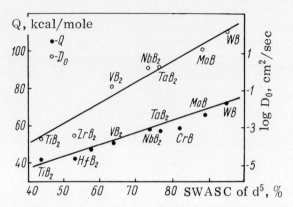

Fig. 57. Activation energy Q and frequency factor D_0 for reactive diffusion of boron in refractory borides plotted as a function of the statistical weight (SWASC) of the d^5 configurations of the metal component.

bound less strongly to the nucleus than in other metals and they can be used to increase effectively the statistical weight of the sp^3 configurations in a crystal by stabilization of the carbon configurations and by participation in the covalent $Me-C$ bonds. The energy E_L decreases along a period or a group because of increase in the probability of the formation of the d^5 configurations by the metal atoms. The energy of the initial excitation and the value of Q both increase along the same direction (Table 33). When we go over from the $Me^{IV}C$ to the $Me^{VI}C$ carbides the electron exchange associated with the rearrangement of the configurations of the metal atoms becomes weaker and the diffusional mobility of carbon in the metal lattice increases with increasing Q. A similar relationship is observed also for the transition metals: the value of E_V increases and that of E_L decreases with increasing principal quantum number of the d valence electrons and therefore, Q increases. Figure 57 shows the dependence of the activation energy and of the frequency factor on the statistical weight of the stable d^5 configurations of the metal in the case of reactive diffusion resulting in the formation of the MeB_2 diboride phases [361]. As in the case of carbides, the values of Q and D_0 increase with increasing statistical weight of the d^5 configurations and the rise is linear (within the limits of the experimental error).

We can quote additional examples but the cases considered above are sufficient to show that the parameters and the mechanism of diffusion can be described qualitatively using the configurational localization model because these parameters and mechanisms are determined directly by the electron structure of matter.

14. Adsorption

According to the modern quantum theory of adsorption, an adsorbed atom exchanges electrons continuously with the adsorbent (substrate) [367]. Hence, the electron structure of the surface affects that energy which is governed by the nature of the chemisorption binding forces between the adsorbed atoms and the atoms on the surface of the substrate. Investigations of the catalytic properties of the transition elements [368, 369] have suggested the possibility of the dependence of these properties on the position of the elements in question in the periodic table. Such a dependence is supported by the experimental results. For example, it has been established [370] that when the energy of binding between a molecule reacting to oxygen and a rare-earth oxide is plotted as a function of the atomic number of the rare earth, it is found that this energy varies in the same way as the binding energy of the metal and oxygen in the oxide and is also a periodic function related to the magnetic moment of rare-earth oxides. This correlation shows that the chemisorption and the volume interactions are similar and that the 4f electrons have the same influence on the adsorption as on other physicochemical properties. A similar conclusion for metallic systems is reported in [371], where it is shown that the chemisorption bonds between the metal substrate and the adsorbate are very similar to the bonds in the bulk state and the binding forces do not differ by more than 10%.

The mixed "chemical" nature of the adsorption binding forces is supported also by the quantum-mechanical broadening of the valence energy levels of the adsorbed particles and by the binding energy which exceeds 1 eV. Quantitative calculations of the adsorption energy of many metallic systems [375, 376] based on this point of view are in good agreement with the experimental results.

The published experimental information on the adsorption properties (adsorption energy) of metal film systems are very

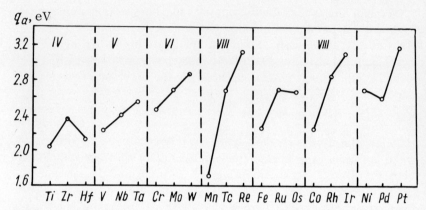

Fig. 58. Adsorption energies of Me−Cs systems under very low coverage (θ ≈ 0) conditions (Me is a d-type transition metal of group IV-VIII).

scarce and they mainly apply to adsorbates on cesium substrates. Recent analysis of the published data, carried out by I. A. Podchernyaeva and V. S. Fomenko, has yielded the following conclusions:

1) The binding energy of the Me−X systems (Me is a IV-VIII group transition metal and X is an alkali or an alkaline-earth metal) definitely tends to increase with increasing principal quantum number of the valence electrons of the adsorbent (substrate) atoms within the same group (Fig. 58).

2) The binding energy of the adsorption systems with a fixed adsorbent (W) and different adsorbates (Cs, Ba, La, Th)

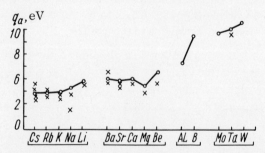

Fig. 59. Adsorption energies of W−X systems under very low coverage (θ ≈ 0) conditions: X) experimental values; O) calculated values.

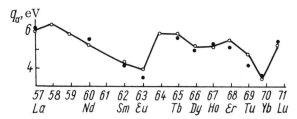

Fig. 60. Energies of adsorption of rare-earth metals (with different number of f electrons) on tungsten under very low coverage ($\theta \approx 0$) conditions: ●) experimental values; O) calculated values.

increases with increasing number of the valence electrons in the adsorbate atoms (Fig. 59).

3) The binding energy of the adsorption systems formed between tungsten and the rare-earth metals shows a definite dependence on the atomic number of the adsorbate (Fig. 60).

Under the chemisorption conditions the interface is regarded as a quantum-mechanical system whose adsorption properties are governed by the nature of the electron interaction between the adsorbent (substrate) and adsorbate atoms. It is obvious that this interaction will increase in strength with increasing difference between the donor properties of the components. In all the systems considered so far the adsorbent surface acts as an acceptor which attracts excess bonding electrons: this give rise to the ionic (electrostatic) component of the bonding. This component is a function of the difference ($I - \varphi$), where I is the ionization potential of the adsorbed atoms and φ is the work function of the substrate. When the values of this difference are small or negative, the ionic components will be strongest [374]. The local overlap of the wave functions in an adsorption system should result in the interaction between the valence electrons of the adsorbate atoms and the d shell of the adsorbent atoms. This interaction is governed by the degree of localization of the valence electrons of the adsorbent and it gives rise to the covalent (nonelectrostatic) component of the bonding.

When the ionic and covalent components occur in the chemisorption bonds, the predominance of one or the other component is governed by the electron structure [373, 374]. Thus, in the case of adsorption of the alkali or alkaline-earth elements on the transition metals the similarity of the values of I of the adsorbed atoms and φ of the substrate gives rise to a strong ionic component which may amount to 40-50% [373]. In the case of metal systems consisting entirely of the transition elements (for example, the adsorption of the rare-earth metals on tungsten), we find that $I > \varphi$ and the ionic component is weak. We may assume that the relative change in the adsorption properties of such systems is primarily due to the covalent component of the bonding because the special nature of the physicochemical properties of the transition elements is governed by the presence of partly filled d and f shells, i.e., by the s−d− f exchange which gives rise to the covalent bonds. The results plotted in Fig. 58 show that the adsorption energy of Cs increases with increasing principal quantum number of the d electrons of the substrate atoms within each group (IV-VIII) of the periodic system.

This may occur for two reasons. An increase in the work function of the adsorbent (substrate) electrons reduces the difference $(I - \varphi)$ and this tends to enhance the ionic component of the bonding. On the other hand, the same increase should strengthen the covalent component of the chemisorption bonds because the degree of localization of the valence electrons of the adsorbent increases with increasing principal quantum number of these electrons (these electrons form the stable d^5 configurations if $n_d \le 5$ and the d^{10} configurations if $n_d > 5$). This stimulates the s−d electron exchange in a metal crystal.

The adsorption of electropositive atoms results in participation of their valence electrons in the indirect d bonds and the participation increases with increasing statistical weight of the stable configurations in the surface layer of the substrate. Thus, we may expect an increase in the adsorption activity of the metals along this direction, which is indeed observed for cesium films and for nearly clean surfaces $(\theta \approx 0)$. In the adsorption of potassium and mercury the relative activity should increase on transition from substrates formed by the IV group metals to those formed by the VIII group metals, which is confirmed for $\theta \approx 0$ in [375] where it is shown that the ability to adsorb cesium increases along the series Ta−W−Re−Ir (Fig. 58). This tendency should be very pronounced for the adsorbates with the predominantly ionic bonding (alkali and alkaline-earth metals). These conclusions are in

agreement with the hypothesis put forward in [376] that the catalytic activity of the transition metals may increase with increasing
proportion of the covalently bound electrons.

It is worth noting the strong fall in the adsorption energy for
the Mn−Cs system (Fig. 58). This anomaly is not accidental but
follows from the special features of the electron structure of manganese which has the $3d^54s^2$ valence shell in the free state. The
relatively low degree of localization of the valence electrons, resulting from the small principal quantum number, and the large
number of the d electrons in free atoms of manganese give rise
to a high density of the collective-state electrons in manganese
crystals and this is why these crystals adsorb weakly. For the
same reasons the electrical resistivity of manganese is high and
the work function is low.

If we consider the adsorption energy q_a as a function of the
electron structure of the adsorbed atoms (Fig. 59), we find that in
the case of adsorption on tungsten the value of q_a increases strongly
with increasing number of the valence electrons of the adsorbed
atoms from 1 for cesium to 4 for thorium. On the other hand, the
value of q_a is practically unaffected when the adsorbates are the
alkali or alkaline-earth metals which always have one or two s
valence electrons, respectively. Obviously, the important factor
is the number of the valence electrons which can interact with the
lattice of the adsorbent. However, when this approach is used in
the interpretation of the experimental results we must allow not
only for the number of the valence electrons of the adsorbate but
also for the structure of the outer electron shells of this substance.
This is illustrated in Fig. 60 which shows the dependence of the
adsorption energy on the atomic number of rare-earth metals adsorbed on polycrystalline tungsten ($\theta \approx 0$). We can see from this
figure that q_a falls from La to Eu and from Ge to Lu. This can
be explained quite easily if we bear in mind that the donor properties of lanthanoids become weaker along the same direction
(this is confirmed, in particular, by the reduction in the higher
valences of the metals in their compounds along the same series
because of increase in the localization of electrons in the f^7 states
as a result of the s → f and d → f transitions). The s → f and
d → f transitions are much less likely in the case of europium
and gadolinium because of the presence of the stable f^7 and f^{14}
configurations in these metals. However, the appearance of one

d electron in gadolinium raises its valence compared with that
of europium and this has a critical influence on the energy characteristics of the adsorption process. Similar behavior should
be observed for Yb ($f^{14}s^2$) and Lu ($f^{14}d^1s^2$), which is confirmed
by the calculations reported in [373].

In the Gd − Lu lanthanoid series the small difference between
the energies of the f^7 and f^{14} configurations (this difference is
smaller than for the f^0 and f^7 states) gives rise to slower fall
of q_a with increasing statistical weight of the f^{14} configurations.
Similar discontinuities are observed in the dependences of the
melting point and the work function on the atomic number of the
rare-earth metals.

It is interesting to compare the adsorption properties of the
rare-earth metals with the corresponding properties of the alkali
and alkaline-earth metals [373].

The numerous published data [377, 378] show that q_a of the
alkali and alkaline-earth elements increases with increasing
work function of the adsorbent. This is due to the stronger ionic
component of the chemisorption bonds formed by these elements
because of the relatively low values of their ionization potentials
[374]. However, in the case of systems consisting solely of the
transition element atoms, in which the covalent bonding predominates, this conclusion is not always correct. Table 34 lists,
for the sake of comparison, the values of the adsorption energy q_a
and of its ionic component H_i for the Me − Cs and Me − La sys-

TABLE 34. Calculated Adsorption Energies q_a of
Me − Cs and Me − La Systems

Adsorbent	Electron configura- tion of free atoms	SWASC of d^5, %	Work function, φ, eV	System	q_a, eV for $\theta \approx 0$	H_i, %
Hf	$5d^2 6s^2$	55	3.53	Hf—La	5.6	0.43
				Hf—Cs	2.14	25.6
Ta	$5d^3 6s^2$	81	4.19	Ta—La	6.0	1.98
				Ta—Cs	2.58	35
W	$5d^4 6s^2$	96	4.62	W—La	6.2	2.75
				W—Cs	2.88	41.6
Re	$5d^5 6s^2$	94	5.10	Re—La	6.2	3.82
				Re—Cs	3.16	48.4

tems, where Me is a transition metal from a long period. An
examination of the values listed in Table 34 shows that the adsorption energy of the systems with the predominantly covalent
bonds (Me − La) is at least twice as large as q_a of the Me − Cs
systems. Moreover, whereas the Me − Cs systems exhibit in all
cases an increase of q_a with increasing φ of the adsorbent, this
does not always apply to the Me − La systems. In fact, q_a (W − La)
hardly differs from q_a (Re − La) although the difference between
the work functions of tungsten and rhenium is about 0.5 eV. This
can be expected for all the rare-earth elements adsorbed on tungsten and rhenium (Fig. 61). The increase in q_a (Me − La) compared with q_a (Me − Cs) can be explained by the larger number of
the valence electrons of lanthanum, which can interact with the
lattice of the adsorbent. The covalent nature of this interaction is
determined by the possible participation of the $5d^16s^2$ electrons of
lanthanum in the s − d electron exchange in the adsorbent crystal.
This is favored by the similarity of the energy states of the valence electrons of the lanthanum and the adsorbent atoms.

When we go over from hafnium to tungsten, we find that the
degree of localization of the valence electrons of the adsorbent
atoms increases (these electrons form the d^5 stable configurations). This stimulates the s − d electron exchange in the metal
crystal and, therefore, the adsorption energy of the lanthanum
atoms increases along the indicated series. This increase in the
electron exchange is reflected in Table 34 by the increase in the
statistical weight of the d^5 configurations. Since the similarity
of the electron structure of the valence subshells of rhenium and

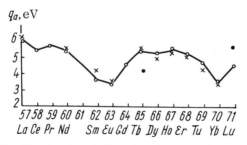

Fig. 61. Energies of adsorption of rare-earth metals
on tungsten (O denotes experimental values and X
calculated values) and on rhenium (● denotes calculated values).

tungsten is responsible for the similar values of the statistical weights of the stable configurations in these metals, it follows that the degree of participation of the valence electrons of the adsorbed atoms in the s−d exchange in the adsorbent should be similar and this is reflected in the similarity of the adsorption energies of the system in question.

15. Surface Tension

Theoretical explanations of the experimental data on surface tension have been given by Ya. I. Frenkel', R. Shuttleworth, Ya. G. Dorfman, A. G. Samoilovich, A. A. Zhukhovitskii, and others. The principal deficiency of the proposed explanations is that they consider a two-dimensional electron gas whose density is assumed to be equal to the density of electrons in the condensed metal phase. Most of the theories deal with the states at absolute zero and they discuss the behavior of a surface layer of metal and of a two-dimensional electron gas. They ignore the individual properties of the metal which are due to the bulk parameters of the positive ion cores.

The dependence of the surface tension σ of liquid transition metals on the degree of occupancy of their d shells (Fig. 62) shows [379] that the surface tension increases with increasing number of the d electrons in the valence shell of free atoms. The higher the principal quantum number of the valence electrons and the higher the stability of the d^5 configurations, the stronger is the surface tension (Fig. 63).

When we go over from vanadium to chromium we find that the surface tension decreases considerably. We have mentioned earlier that the statistical weight of the d^5 configurations of chromium is less than that of vanadium, beginning from temperatures of the order of the melting point. Consequently, the surface tension of chromium is less than that of vanadium. The lower surface tension of the group VI metals, compared with the group V elements is observed also for metals in the second long period and even in the third long period (tungsten). Another special feature which distinguishes the surface tension from the other high-temperature properties determined by the atomic binding in the condensed state are the high values of σ of the elements whose free atoms have six or seven electrons in the d shell. This applies particularly to osmium and iridium in the third long pe-

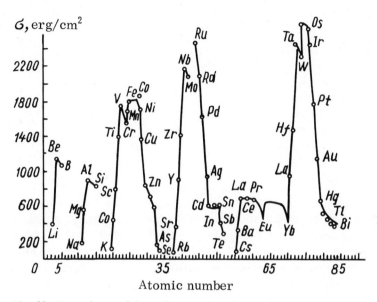

Fig. 62. Dependence of the surface tension of elements on the degree of completion of their valence shells in the various periods.

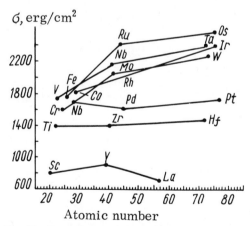

Fig. 63. Dependence of the surface tension of the d-type transition metals on their atomic number (different groups are shown separately).

riod. A similar relationship applies to the low-temperature mechanical and thermodynamic properties such as Young's modulus (Fig. 50), or to the Debye temperature (Table 8). It is explained by the partial "expulsion" of the excess of d electrons into the sp states which results in an additional contribution of these electrons to the binding and produces an associated increase in the statistical weight of the stable d^5 configurations. At high temperatures this effect is suppressed by the d localization, which increases the statistical weight of the d^{10} configurations. It is likely that disturbed configurations of this type remain on the surface of the condensed phase and this leads to an increase in the surface tension.

Palladium and platinum exhibit high statistical weights of the d^{10} configurations in the condensed phase and, therefore, these elements are characterized by low surface tension. The high-temperature features of the configuration spectra determine also the relationships governing variation of the surface tension in the Fe−Co−Ni triad.

Lanthanoids, whose free atoms have unfilled f shells, obey the general dependence of the surface tension on the atomic number. As in the case of the d-type metals the bonds in the f-type transition elements are formed mainly by the d electrons. The tendency of the f shell to reach the stable f^7 state reduces the statistical weight of the d^5 configurations from lanthanum to samarium with a specially large drop at europium for which the statistical weight of the d^5 configurations and the fraction of the s electrons participating in the bonding are exceptionally low because of the formation of the stable f^7 configuration.

Similar behavior is observed also in the second halves of the long periods: the tendency for the statistical weight of the stable f^{14} configurations to increase rises between gadolinium and thulium. Like europium, ytterbium is characterized by the greatest fall in the number of the bonding electrons because of the transfer of some of them to the f shell and the formation of the f^7 stable configurations. Since the tendency to increase the statistical weight of the f^0 configurations in a crystal implies a simultaneous increase in the density of the bonding electrons and of the statistical weight of the d^5 configurations, whereas the tendency to form the stable f^7 configurations has the opposite effect, the values of surface tension in the first halves of the long periods are lower than in the second halves.

16. Contact Phenomena

The interaction between two or more bodies in contact determines the strength of a joint produced by cold welding [380] or diffusion bonding [381], the wear resistance of parts subject to friction [382] and of cutting tools [383-384], the performance of an abbrasive disk in grinding [385-386], etc.

Adhesive interaction results from the contact between two bodies in which intermingling may occur at the atomic and electronic level along the interface [387]. In the initial stage, the two surfaces are prepared and brought sufficiently close together for the atomic interaction to occur. This may be achieved by the simultaneous plastic deformation of the two bodies or by surface migration of the atoms. The final stage is dominated by the electron interaction processes, which are determined by the electron structure of the substances in contact.

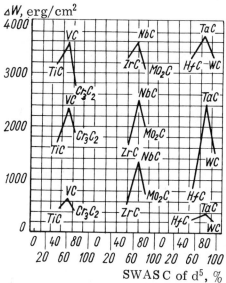

Fig. 64. Work of adhesion of liquid Al, Fe, and Cu in contact with carbides of the d-type transition metals.

It follows from general considerations that adhesion bonding should result in the formation of stable configurations between atoms of the pair of the substances in contact. Such formation of the stable configurations can result from electron exchange between the atoms. If the two substances in question have high values of the statistical weight of the stable configurations, the adhesion may be quite weak. Figure 64 shows the dependences of the work of adhesion of liquid La, Fe, and Cu to carbides — calculated from the relationship $W_A = \sigma_L (1 + \cos \Theta)$ — on the statistical weight of the stable d^5 configurations of the metal components of the carbides [386]; here, W_A is the work of adhesion, σ_L is the surface tension of the liquid metal, and Θ is the wetting angle. Similar results are obtained by plotting the dependences of the work of adhesion on the number of electrons in the d shells of the metal atoms in the carbides. The relationships obtained indicate that when carbides are wetted by metals of different nature, the work of adhesion W_A varies always in the same manner with the electron structure of the metal component of the substrate. Similar relationships are obtained also for the dependences of other properties of the carbides on the number of electrons in the d shells of the metal atoms.

The group IV carbides are characterized by relatively low values of the work of adhesion W_A . This may be due to the fact that, in addition to low values of the statistical weights of the d^5 configurations, they are characterized by relatively high values of the statistical weights of the sp^3 states, which reduce the fraction of the d collective-state electrons (intermediate spectrum), capable of participating in the exchange between surfaces.

Since the carbides of the group VI metals have high statistical weights of the d^5 configurations but few delocalized d electrons capable of participating in the interaction between the surfaces, the work of adhesion of metals to these carbides is small. The optimal values of the work of adhesion of metals to carbides are obtained for the carbides of the group V metals in which the number of unsaturated bonds is sufficiently high, i.e., the statistical weight of the intermediate d^n configurations is high.

The same approach can be used in the elucidation of the dependences of the work of adhesion on the atomic number of the transition metals in a given period or group. It must be stressed that just as the strength of bonds between atoms in a lattice governs many properties, the contact interaction between surfaces also determines many properties and processes. Diffusion bonding is an example of a process which can be described in terms of the contact interaction between the surfaces [388].

In general, diffusion bonding can be described as consisting of two stages: the initial adhesion bonding stage and the final diffusional migration of the substances in contact. In the case of systems with high statistical weights of the stable electron configurations, we may expect an increase in the temperature of the diffusion bonding because of the difficulty of exciting the stable electron configurations in the initial stage. The second stage of bonding involves diffusion which results in the matching of the structures and the dispersal of microdefects, i.e., processes which determine the quality of the bonded joint. The diffusion processes and their energy characteristics also depend on the statistical weight and the stability of the electron configurations.

Thus, both stages of diffusion bonding are governed by the same parameters and are basically of the same origin. The only difference is that during the adhesion stage the action takes in the surface layers of atoms in which the statistical weight of the stable electron configurations is less than in the bulk and, therefore, the process of adhesion should be faster and should require less excitation (other conditions being equal) than the subsequent diffusion process. The diffusion method has been used successfully to bond such d-type transition metals as nickel, titanium, niobium, molybdenum, and tungsten. We now know the optimal temperatures, pressures, and durations of heating which are necessary to ensure rigid and high-quality joints whose strength is equal to that of the original metal, i.e., we know the conditions under which the physical interface between the joint parts can effectively disappear. These optimal parameters are listed in Table 35.

The optimal bonding temperatures increases monotonically with increasing statistical weight of the d^5 configurations of the metals being bonded. However, the ratio of the bonding temperature to the melting point remains practically constant, which is due to similar dependences of the two temperatures of d-type elements on the statistical weight of the stable electron configurations. Attempts to establish a correlation between the degree of

TABLE 35. Optimal Conditions
for Diffusion Bonding of
Transition Metals

Metal	Bonding temp., °C	Duration, min	Pressure, kg/mm²
Ni	1000	20	1.5
Ti	1000	5	0.5
Nb	1200	15	1.5
Mo	1500	15	1.5
W	1800	10	1.0

adhesion of bonded pairs on their hardness [389] or on the hardness and the melting point [390-391] are equivalent only to the extent to which the melting point and the hardness reflect the excitation of the stable electron configurations. The correlation with the hardness is less satisfactory because the hardness depends strongly on the low-temperature statistical weight of the stable configurations.

Diffusion bonding of refractory transition metals with refractory carbides, borides, and nitrides is of special interest. The delocalized electrons of the transition metal in such refractory compounds are partly transferred to the atoms of the nonmetal (for example, carbon), and this increases the statistical weight of the $s^x p^y$ configurations of the nonmetal (these configurations are formed by the localized fraction of the valence electrons of the nonmetal). When a transition metal is brought into contact with a refractory compound of this type, the adhesion and the subsequent diffusion are governed primarily by such rearrangements and electron exchange which tend to increase the statistical weights and stability of the electron configurations above the level found in the original substances. It is necessary to distinguish two main cases within the same class of compounds: $MeC - Me$ (the bonding of a metal to its own carbide) and $MeC - Me'$ (the bonding of a metal to a carbide of a different metal).

In the first case, the bonding may result mainly in the formation of the lower carbide phases if such exist. It follows from the data in Table 36 that such phases are indeed formed in the bonding of niobium, tantalum, and tungsten to their higher carbides. We can expect an increase in the statistical weight of the

TABLE 36. Conditions for Bonding of Carbides
to Refractory Metals

Bonded pair	Bonding temp., °C	Duration, min	Pressure, kg/mm²	Phase formed
TiC—Mo	1700	10	0.5	Mo₂C
ZrC—Nb	1400	10	1.5	NbC
ZrC—Ta	2000	10	0.5	80% TaC + 20% ZrC
ZrC—Mo	1500	10	1.5	Solid solution of C in Mo
ZrC — W	1800	10	1.5	Solid solution of C in W
NbC—Nb	1600	10	0.5	Nb₂C
NbC—Ta	1700	10	0.5	60% NbC + 40% TaC
NbC—Mo	1800	10	0.5	Solid solution of C in Mo
NbC — W	1800	10	0.5	70% NbC +30% WC
TaC—Nb	1200	10	0.5	Solid solution of C in Nb
TaC—Ta	1900	10	0.5	Ta₂C
TaC—Mo	1600	10	0.5	Ternary phase Ta—Mo—C
TaC — W	2000	10	0.5	Ternary phase Ta—W—C
Mo₂C—Mo	1400	10	0.5	Mo₂C—Mo
Mo₂C—W	1500	10	0.5	Solid solution of C in W
WC—Mo	1850	10	0.5	Mo₂C and ternary phase Mo—W—C
WC—W	1900	10	0.5	W₂C

electron configurations of the transition metal and the carbon atoms and some reduction in the number of the delocalized electrons relative to the numbers existing originally in the metal and carbon. The bonding temperature should be governed primarily by the energy necessary for the excitation of the stable configurations in the carbide and in the metal. This applies to the systems NbC−Nb, TaC−Ta, and WC−W. In the Mo₂C−Mo system, in which the formation of a lower carbide is impossible, the bonding may primarily be due to the transfer of some of the delocalized electrons from the atoms in molybdenum carbide to the atoms in

metallic molybdenum. Such transfer is easy because no fundamental changes in the structure are required and that is why the
bonding temperature of this pair is relatively low (1400°C).

In the second case (the bonding of a metal to a carbide of a
different metal) the bonding temperature depends strongly on the
ease of excitation of the stable configurations of atoms in the two
materials but the degree of stability of the phases formed as a
result of bonding is of greater importance. It follows from the
results in Table 36 that when the statistical weight and the stability of the electron configurations of the pure metal increase, solid
solutions of carbon in both metals rather than new carbides are
formed.

The increasing difficulty to destroy the carbide phase MeC
and to form carbides of the second metal Me' is due to the increase in the fraction of the delocalized electrons in the reaction
MeC + Me' = Me'C + Me(C). Consequently, this reaction is unlikely to occur. On the other hand, the increase in the statistical
weight of the stable configurations in the carbide phase, i.e., the
reduction of the difference between the statistical weights of the
stable configurations in the carbide and the second metal, increases the probability of formation of new carbide phases during
welding. This follows, for example, from a comparison of the
nature of the phases formed in the following series: $ZrC-Me$
(increase in the density of the delocalized electrons as a result
of formation of MeC phases) and $TaC-Me$, Mo_2C-Me, $WC-Me$
(reduction in the fraction of the delocalized electrons as a result
of formation of MeC phases).

Experimental investigations have been made of the possibility
of the diffusion bonding of metals to boron carbonitride BNC, in which
all the atoms have high statistical weights of the sp^3 configurations (an atom of nitrogen acquires the sp^3 configuration by the
transfer of one electron to the atom of boron which then also assumes the sp^3 configuration; the s^2p^2 configuration of carbon
changes to the sp^3 configuration because of the s → p transitions).
These investigations have indicated that no interaction occurs between boron carbonitride and niobium, tantalum, molybdenum, and
tungsten right up to the melting points of these metals. The high
statistical weight of the sp^3 configurations of diamond is responsible for the poor wettability of this substance when in contact

with molten metals and alloys. This applies also to other non-metals and nonmetallic compounds with high statistical weights of the stable electron configurations. Obviously, the wettability and the adhesion in diffusion bonding can be increased mainly by disturbance of the stable configurations or by formation of compounds with higher statistical weights of the stable electron configurations. Alternately, the stable configurations in the nonmetal can be disturbed by introducing impurities. This is why the wettability of diamond by molten metals generally increases with increasing impurity concentration in this material.

The electron interaction of the boundary between two bodies is also responsible for the "contact fusion" [392-396] in which a liquid phase forms at temperatures below the melting point of the more easily fusible component. This phenomenon is observed for systems with a minimum in the fusion phase diagram.

The contact fusion is governed primarily by the changes in the nature of the atomic bonding in a system as a result of diffusion of atoms of one component into the lattice of the other component. Such diffusion always results in an increase of the statistical weight of the stable electron configurations and a reduction of the free energy. If, additionally, the statistical weight of the half-filled bonding configurations decreases and the density of the collective-state electrons or the statistical weight of the atomic stable states of one of the components increases (the total statistical weight of the stable configurations in the system also increases), we find that the reduction in the strength of the atomic binding of the surface particles is manifested as a fall of the interface surface energy which gives rise to contact fusion. We shall follow the work of Shebzukhov and Savintsev [392] who considered contact fusion between transition metals within the framework of the configurational localization model. We shall discuss the relationships governing this process in the case of vanadium—transition metal systems (Table 37) and more specifically the $V-Ti$ system.

When vanadium atoms diffuse into titanium, they tend to capture electrons in order to increase the statistical weight of their own d^5 configurations, which reduces the statistical weight of the d^5 states in the titanium matrix. The total free energy of the system decreases (because of increase in the statistical weight of the d^5

TABLE 37. Contact Fusion
Temperature T_c of Systems
Formed between
Vanadium and Transition
Metals and Atomic
Concentration X of
Transition Metal in Liquid
in Contact Zone

Metal	T_c, °K	X, at.%
Ti	1893	71.3
Cr	2023	69.5
Zr	1503	55.0
Nb	2083	22.8
Hf	1720	50.0
W	1923	4.5
Y	1738	88.0
Fe	1741	77.0
Co	1521	57.5
Ni	1476	49.5
Pd	1613	39.0

configurations of the vanadium component) and the energy of the
bonds in the titanium matrix also decreases. Thus, the energy of
the bonds in the titanium-base solid solutions formed in this way
is less than in the pure metal and therefore contact fusion occurs.
Other mechanical properties also decrease but the resistivity increases [397] because the phonon component of the electron scattering becomes stronger. The reduction in the statistical weight
of the d^5 configurations of titanium, resulting from the addition
of vanadium, is accompanied by the formation of the d^5 configurations of vanadium. When the concentration of vanadium is increased, the statistical weight of the d^5 configurations of the impurity increases and at some concentration the formation of these
configurations becomes more important than the increase in the
statistical weight of d^0 configurations of titanium. Beginning from
this configuration, we should observe a rise of the melting point
and a fall of the electrical resistivity with increasing concentration of vanadium, which is in agreement with the phase diagram
of this system [398] and with the results of studies of the dependence of the electrical conductivity of titanium—vanadium alloys
on their composition [397]. A similar situation arises also in the
Zr—V and Hf—V systems in which a considerable contribution to
the bonding is made by the s^2 configurations. Since the stability of

the s^2 configurations decreases from titanium to zirconium and the probability of the transfer of the s electrons to the d^5 states of vanadium increases along the same direction, we find that the contact fusion temperature of the Zr−V system is lower than that of the Ti−V system. This temperature increases somewhat for the Hf−V system because a considerable contribution to the bonding of the group IV metals is made not only by the s^2 configurations but also by the d^5 states whose stability rises strongly with increasing principal quantum number, masking the fall of the contribution of the s^2 configurations of hafnium.

At contact fusion temperatures, which are of the order of 1500-2000°K, the statistical weight of the d^5 configurations of chromium is lower than that of vanadium and, therefore, electrons may be transferred from the chromium atoms to the d^5 configurations of vanadium. This transfer is less pronounced in chromium than in titanium, zirconium, or hafnium and therefore the statistical weight of the remaining d^5 configurations of chromium is considerably higher than the corresponding parameters of titanium. The contact fusion temperature increases accordingly: this temperature is 1893°K for the Ti−V system and 2023°K for the Cr−V system. The statistical weight of the d^5 configurations of molybdenum decreases (like that of chromium but much more slowly) when the temperature is increased and, probably, it approaches the corresponding statistical weight of vanadium. Therefore, the addition of vanadium to molybdenum cannot result in a significant reduction in the statistical weight of the stable electron configurations of either of the components and no contact fusion is observed between vanadium and molybdenum. This approximate equality of the statistical weight of the d^5 states of the components applies also to the Ta−V system.

The statistical weight of the d^5 configurations of niobium increases with increasing temperature and it may rise above the corresponding statistical weight of molybdenum. This weight is even higher for tungsten because of the high stability of the d^5 configurations, which are weakly disturbed by the rising temperature. Therefore, the reverse process is observed in these systems: vanadium atoms give up electrons which are used to increase the statistical weight of the stable configurations of niobium or tungsten. According to [392], a solid solution is formed with vanadium as the base and the reduction in the statistical weight of the stable d^5 configurations in the V — W system is greater than in the V — Nb system. Therefore, the contact fusion temperature in the V — W system is lower than in the V — Nb system.

Each of the elements in the iron triad exhibits contact fusion with vanadium because of attraction of the vanadium electrons into the stable d^{10} configurations. This attraction effect is exhibited more strongly by nickel so that the reduction in the statistical weight of the d^5 configurations is greatest for vanadium in contact with nickel. Consequently, the contact fusion temperature decreases along the series Fe — Co — Ni. This temperature increases slightly when nickel is replaced with palladium but no contact fusion is observed for ruthenium and rhodium. It is also absent when scandium, lanthanum, manganese, and rhenium are brought into contact with vanadium. As mentioned earlier, the tendency for the formation of the stable d^5 configurations in ruthenium, rhodium, osmium, and iridium results in the "excess" d electrons becoming delocalized in the sp levels from which they can be transferred quite easily to vanadium. Therefore, the statistical weight of the stable d^5 configurations of vanadium increases. The same delocalization effect is manifested in some of the other systems mentioned above because of the tendency of their atoms to retain the d^0 (in the case of Sc and La) or the d^5 (in the case of Mn and Fe) configurations. Therefore, the energy of the atomic bonds in solid solutions based on these elements is higher than the energy of bonds in the separate components and contact fusion does not occur.

The importance of the stable configurations in contact processes at lower temperatures is demonstrated in [400] where the electrochemical treatment (a form of electro-erosion treatment) of heterogeneous pairs of transition metals is considered. As in contact fusion, the electron exchange intensity and the different tendencies to shift the electron density in the contact zone result in adhesion and transport of matter from the samples to the electrodes (positive wear) or from the electrodes to the samples (negative wear).

Titanium in the condensed state has a relatively low statistical weight of the stable d^5 configurations and therefore it is possible to increase this weight by using titanium as the electrode material in the electromechanical treatment of transition metals.

This increase results from the transfer of some of the valence
electrons from titanium to the other metal. Such a transfer shifts
considerably the electron density, which determines the nature of
the adhesive interaction in the contact zone, as well as seizure
and transfer of titanium to the metal being treated. Obviously,
in this situation the wear of the electrode should be negative, i.e.,
the transfer should take place from the electrode to the material
being treated, which is indeed confirmed by investigations of the
Ti−Zr, Ti−Nb, and Ti−W pairs. The intensive wear of titanium
electrodes is illustrated by the data listed in Table 38. The transfer of matter from titanium to zirconium results in some increase
in the statistical weight and the stability of the d^5 electron configurations. In particular, the energy considerations favor the
transfer of electrons from titanium to zirconium with a corresponding shift of the electron density in the contact zone. For the
same reasons the use of zirconium electrodes in treatment of
titanium results in seizure and transfer of matter from titanium
to zirconium (positive wear). Conversely, in the Zr−Fe and Zr−Ni
pairs the zirconium should transfer electrons which are used to
complete the d^{10} configurations of iron and nickel and therefore the
wear of the electrodes should be negative, which is confirmed experimentally.

The higher value of the statistical weight of the d^5 configurations in niobium compared with zirconium and the tendency for
this weight to increase in the Ti−Nb system leads to the preferential formation of the d^5 configurations in the contact layer of

TABLE 38. Wear of Electrodes ($\Delta_W \times 10^{-3}$, mm^3)

Electrode	Worked material						
	Ti	Zr	Nb	Mo	W	Fe	Ni
Ti	+72.8	−6.1	−6.0	—	−4.3	—	—
Zr	+62.5	+138.2	—	—	—	−5.1	−1.3
Nb	+34.8	—	+5.6	−7.5	—	−1.9	−0.8
Mo	—	—	+2.1	−14.7	—	+0.5	−0.2
W	+6.7	—	—	—	−8.7	—	−0.1
Fe	—	+13.5	−1.9	−11.0	—	−1.6	−5.1
Ni	—	+11.5	+3.0	−10.1	−2.6	+3.2	−0.1

Note. The wear of the electrodes is calculated from the formula $\Delta_W =$
$(\sigma_1 - \sigma_2)/\rho$, where σ_1 is the weight of the electrode before the working; σ
is the weight of the electrode after the working; ρ is the density of the material being worked under constant conditions.

niobium atoms: this also leads to seizure and transfer of matter from the electrode to the material being worked. Similar wear should also be observed in the $Zr-Nb$, $Ti-V$, $Ti-Ta$, and $Zr-Ta$ pairs.

The group VI metals have even higher statistical weights of the d^5 configurations and a stronger tendency to increase these weights. Therefore, when molybdenum or tungsten are used in treatment of other materials, in most cases the material being treated is transferred to the electrode. Different results are obtained only in the treatment of nickel for which wear of tungsten and molybdenum electrodes has been reported. It is interesting to note that in the reverse process of treatment of the group VI metals with nickel, it is found that the nickel electrode suffers wear, i.e., in both cases no seizure has been reported. As noted earlier, nickel, molybdenum, and tungsten exhibit the strongest tendency to increase the statistical weight of the stable (d^{10} for nickel and d^5 for molybdenum and tungsten) configurations. Therefore, any redistribution of electrons in systems of this kind would unavoidably result in the destruction of the stable configurations in one of the components and this is not favored by the energy considerations. Consequently, the electron density in the contact layer cannot shift significantly in any one direction and seizure does not occur. In such cases the electrodes and the samples suffer wear during electromechanical treatment and the wear of tungsten and molybdenum is considerably less (Table 38) because of the bonding nature of the stable configurations formed in these metals (the strength of bonds in tungsten and molybdenum is higher than in nickel).

Similar studies have been carried out on iron and nickel and it has been found that all the materials being treated are deposited on nickel electrodes (with the sole exceptions of tungsten and molybdenum). This is evidently due to the very strong tendency of nickel to complete the stable d^{10} configurations at the expense of the electrons in the metal being treated. In the case of iron the statistical weights of the d^5 and the d^{10} configurations are approximately equal (54 and 46%) so that transfer of the material being treated (if it has a low statistical weight of the stable configurations) to the nickel electrode as well as pronounced wear of the electrode (in the opposite case) is possible. The results obtained indicate that zirconium tends to collect on iron electrodes whereas the treatment of niobium, molybdenum, and nickel results in wear of the iron electrode, which is strongest for the Fe−Mo system.

It is interesting to compare the principal relationships which are observed when the electrode and the material being treated are interchanged. The differences between the nature and the intensity of the processes that occur can be explained by the effect of temperature on the electron structure because the electrode material operates at temperatures much higher than the rotating sample. Therefore, an analysis of these relationships can be made bearing in mind the basic hypothesis according to which the rise in temperature reduces the statistical weight of the stable configurations in metals with the highest localization of the valence electrons (group VI and VIIIC metals) and increases this weight in other cases. Obviously, in most cases, the temperature dependence of the statistical weight is not sufficiently strong to alter the nature of wear. Therefore, in both cases, i.e., in systems A−B and B−A, matter is transferred to the same material but the rate of this transfer is different. For example, in the case of a system formed by titanium and some metal A, the matter is always transferred to the metal A. However, in the Ti−A case the temperature of titanium is higher than in the A−Ti case and, therefore, the statistical weight of the d^5 stable configurations is higher. Consequently, the transfer of the "hot" titanium should be less than of the "cold" titanium, as shown clearly in Table 38. Naturally, the statistical weight of the stable configurations in the metal A also changes with temperature. However, the temperature dependence of the statistical weight of the d^5 states in titanium is so strong (in particular, the rapid rise of the statistical weight of the d^5 states with increasing temperature causes the $\alpha \rightarrow \beta$ transition in titanium) that it dominates the wear processes. These comments apply also to the Zr−A systems because the statistical weight of the d^5 states of zirconium varies rapidly with temperature, as indicated by the $\alpha \rightarrow \beta$ transition in this metal and its "anomalous" behavior in diffusion.

The statistical weight of the d^5 configurations and the degree of localization of the d electrons in niobium and iron vary less rapidly with temperature than in titanium. Therefore, an interchange of the electrode and the material being treated in the Fe−Nb system should not alter significantly the rate of transfer of matter, which is supported by the results given in Table 38: according to this table the values of wear in both cases are equal (within the limits of the experimental error). On the other hand, the statistical weight of the d^5 configurations in molybdenum decreases with increasing temperature and therefore the change from Fe−Mo to the Mo−Fe system reduces considerably the transfer of iron to molybdenum. A similar effect is observed in the Mo−Nb system and in the treatment of nickel with an iron electrode. In the case of treatment of iron with a nickel electrode the statistical weight of the d^{10} configurations in nickel decreases with increasing temperature in the contact zone and this reduces the possibility of seizure of nickel with iron. The Nb−Ni system is similar to the Mo−Ni system, particularly at high temperatures when the statistical weight of the d^5 configurations in niobium approaches the corresponding weight of the d^5 configurations in molybdenum. If niobium is being treated the statistical weight of the d^5 configurations in this metal decreases because the temperature of the material being treated is lower than that of the electrode. The electron density is displaced more strongly toward nickel and this tends to result in stronger seizure of nickel with niobium. Thus, the reported investigations indicate that the processes occurring in the contact zone during electromechanical treatment cause such changes and such electron transfer which tend to reduce the free energy of the system as a whole.

An analysis of the contact phenomena from the point of view of the electron processes occurring at the interface of like or unlike materials explains also the many relationships observed in the friction of homoatomic and heteroatomic pairs.

The characteristics which determine the antifriction properties of materials include their strength, brittleness, plasticity, adhesive power, as well as sintering and oxidation tendencies. If we use the configurational model, we can show that the condition for strong interaction and, therefore, heavy wear in friction is the exchange of electrons between the atoms of the rubbing pairs, which results in the formation of stable electron configurations. Figure 65 shows the dependence of the friction coefficient on the rate of linear wear of homopairs of transition metals immersed in liquid nitrogen [401]. An analysis of the data in Fig. 65 shows that the friction coefficient increases with increasing statistical weight of the intermediate electron configurations in the metals forming a rubbing pair. Therefore, the curve in Fig.

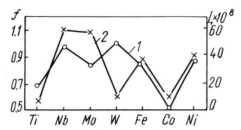

Fig. 65. Friction coefficient (1) and rate of wear
(2) of homoatomic pairs in liquid nitrogen.

65 is closely correlated with the dependence of the electronic
specific heat on the number of electrons in the d shell (Fig. 11).
The intermediate configurations have a natural tendency to accept or donate electrons in such a way as to form stable states.
Consequently, intensive electron exchange with configurations in
neighboring atoms is stimulated. These intermediate configurations are formed in the outer atoms on the surfaces in contact,
and they are responsible for the seizure of the surfaces on the
subatomic scale. The rate of wear is minimal in those cases
when the atoms in the metals forming a rubbing pair have a high
proportion of free sp electrons. The bonds formed during friction
are weak and they are easily destroyed by the relative motion of
the rubbing surfaces. A reduction in the rate of wear is also observed when the probability of electron exchange is very low because of the high statistical weight of the d^5 configurations capable of forming "self-closed" covalent bonds. In this case, the
number of free bonds decreases, which tends to reduce the linear
wear rate.

A regular increase in the friction coefficient is observed
when the transition metals are rubbed on nitrided steel. The presence of nitrogen in the surface layer plays a definite role in this
increase, because the interaction between the rubbing surfaces is
enhanced as a result of the increase in the statistical weight of
the nitrogen atoms in the contact layer, which tend to form the
stable s^2p^6 states. A reduction in the rate of wear is observed
when the metal atoms form the most stable configurations, which
are difficult to disturb. This is why the rate of linear wear decreases rapidly during the rubbing of the cubic transition metals
(i.e., metals based on stable configurations) on nitrided steel,
whereas this rate remains practically constant during the rubbing

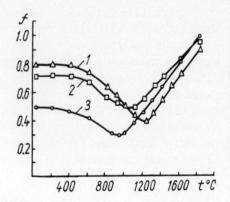

Fig. 66. Temperature dependences of the friction coefficient of homoatomic pairs of refractory carbides: 1) HfC—HfC; 2) ZrC—ZrC; 3) TiC—TiC.

of the hexagonal metals (i.e., metals whose weight of the intermediate spectra is high and which have a high proportion of the delocalized sp electrons).

Investigations of the antifriction properties of refractory compounds [402] have established that:

1) the friction coefficient of the transition-metal carbides increases with increasing atomic number of the metal in each period of Mendeleev's table;

2) if we compare the low-temperature (measured between room temperature and 800-1000°C) friction coefficient of these carbides, we find that the coefficient increases in a regular manner with increasing atomic number of the metal component within each group, whereas the high-temperature coefficients show an opposite tendency;

3) studies of the influence of porosity on the friction of otherwise identical samples of TiC and SrC, show that the friction coefficient increases with increasing porosity;

4) the friction coefficient is almost independent of the temperature in the range 200-400°C but at higher temperatures it exhibits a monotonic rise (Fig. 66), which is preceded by a minimum located at 950, 1100, and 1250°C for the TiC—TiC, ZrC—ZrC, and HfC—HfC pairs, respectively.

The temperature at which the friction coefficient has its minimum value increases with increasing atomic number also in the case of carbides of the group V and group VI metals. Further elevation of the temperature results in a rise of the friction coefficient.

Apart from various macrofactors, the fusion and adhesion are the processes which govern the basic variations of the friction coefficient [403] and they are determined by the same electron exchange mechanism. Thus, the antifriction properties of carbides can be considered from the point of view of the configurational model. When this model is applied to the friction coefficient, we find that the first two experimental dependences mentioned above are simple consequences of the increase in the statistical weight of the intermediate spectra along periods and groups of the metal components (this is supported also by a rise in the magnetic susceptibility and a reduction of the microbrittleness), which occur simultaneously with increasing probability of the formation of stable d^5 configurations because of the weakening of the $Me-X$ bonds.

When the temperature is raised, the intermediate configurations are disturbed and the statistical weight of the d^5 configurations increases, which is why refractory compounds become harder [404] and their friction coefficient become smaller.

Further increase in the temperature tends to raise strongly the statistical weight of the intermediate $s^x p^y$ configurations, which masks the reduction in the weight of the intermediate d^n configurations. The friction coefficient begins to rise and this is most pronounced in the case of TiC, for which the statistical weight of the sp^3 configurations is highest at low temperatures. For this reason, the temperature corresponding to the friction coefficient minimum is lowest for TiC and shifts toward higher values along the series $TiC-ZrC-HfC$.

At high temperatures, the strongest rise in the intermediate $s^x p^y$ spectra in TiC is responsible for the highest value of the friction coefficient of this compound, which decreases when we go over to hafnium carbide. This change in the variation of the friction coefficient in the same group of metals is similar to that observed for the high-temperature microhardness (Fig. 49).

The rise in the friction coefficient with increasing porosity is mainly due to the increase in the coefficient caused by the large number of pore edges rather than to the damage or wear of porous samples. Very active atoms with distrubed electron configurations are concentrated at these edges and these atoms show the strongest tendency for electron exchange. Therefore, the "mechanical" factor of the presence of pores represents effectively the electron factor in a less explicit form. Similarly, until quite recently the grinding process was attributed to "mechanical" processes and the suitability of abrasive materials under a given set of grinding conditions were usually established roughly simply on the basis of physicomechanical processes. These views were revised first in connection with the new abrasive materials, particularly boron carbide. In spite of the fact that this compound is harder than corundum or silicon carbide, disks made of this material are rapidly blunted and are not very efficient.

Experimental investigations [404] have demonstrated that boron carbide becomes oxidized in the presence of atmospheric oxygen and under the influence of the heat evolved in grinding. Consequently, the surface layers of an abrasive tool lose carbon and this alters drastically the properties of such tools. Thus, the chemical interaction of boron carbide with the material being ground is the decisive factor which determines its unsuitability as an abrasive material. It is now known [405] that the oxidation of abrasive materials is not limited to boron carbide but is encountered also in other substances. For example, when titanium and its alloys are ground with disks made of white "electrocorundum," it is found that defects formed in the surface layer of parts made of this metal or its alloys reduce their strength by 25-40% below the initial value. Other studies [406] have established that when titanium alloys are ground, the wear due to shearing or crushing is practically negligible and the interaction between the abrasive and the material being ground is frequently due to diffusion. Clearly, these undesirable phenomena are governed primarily by the nature of the electron structure of the abrasive and the material being abraded. Titanium has a low statistical weight of the d^5 stable configurations and a considerable proportion of the delocalized electrons, whereas abrasives are covalent compounds formed by the sp elements with low or high values of the statistical weights of the sp^3 and s^2p^6 configurations.

When titanium comes into contact with such abrasive, it is energetically likely that the delocalized electrons of titanium will be transferred mostly to the atoms of the abrasive and that this will result in an increase in the statistical weights of the sp^3 and s^2p^6 configurations (the stabilization of these configurations). This is manifested by the transfer of titanium to the grains of the abrasive and a considerable change in the strength of titanium.

It is worth mentioning that this effect should be even stronger in the grinding of scandium, yttrium, and lanthanum.

It follows from our discussions that in order to reduce the "chemical interaction" between a metal and an abrasive, the latter should have "self-closed" bonds, i.e., it should have a high value of the statistical weight of the stable configurations and a low value of the statistical weight of the intermediate configurations (the latter participate actively in the exchange between the surfaces in contact). Moreover, an abrasive should be a hard material and therefore it is reasonable to search for abrasive materials among refractory compounds formed by the transition metals and non-metals in the second period, for which the stability of the sp^3 configurations is highest.

We can see that among carbides the necessary requirements are satisfied by the titanium, zirconium, and hafnium carbides which have the highest hardness and the most completely "closed" bonds, i.e., they are characterized by a low value of the statistical weight of the intermediate configurations. This weight should increase in carbides of the group V and VI metals because of the reduction in the statistical weight of the sp^3 configurations and the weakening of the $Me-C$ bonds which is not compensated (with the exception of tungsten carbide) by some enhancement of the $Me-Me$ interaction (the stability of the sp^3 configurations is higher than that of the d^5 states). The same factors tend to enhance the interaction between titanium and an abrasive, i.e., the refractory carbides Me^VC and $Me^{VI}C$ are not very promising as abrasives for this metal. On the other hand, borides are characterized by a higher value of statistical weight of the less stable sp^2 configurations. Therefore, the statistical weight of the d^5 states in borides determines (to a greater extent than in carbides) the localization of electrons, the statistical weight of the intermediate configurations, and the strength of these materials.

An analysis of this type carried out on the basis of the configuration model [407] has shown that the most suitable abrasives for the grinding of titanium alloys would be refractory compounds based on zirconium carbide and tungsten boride. The results of an experimental check of these predictions are plotted in Fig. 67 (for samples ground with a tape carrying zirconium carbide grains) and in Fig. 68 (for samples ground with a tape carrying tungsten boride grains). Figure 69 shows, for the sake of comparison, the

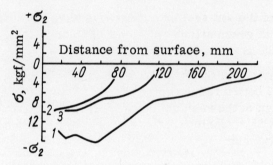

Fig. 67. Distribution of the stresses resulting from the grinding of the titanium alloy VTZ-1 with zirconium carbide.

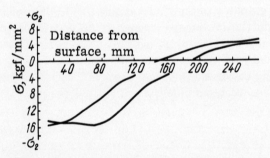

Fig. 68. Distribution of the stresses resulting from the grinding of the titanium alloy VTZ-1 with tungsten boride.

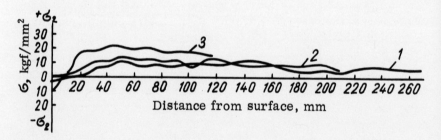

Fig. 69. Distribution of the stresses resulting from the grinding of the titanium alloy VTZ-1 with green silicon carbide.

distribution of the stresses in a sample ground under the same
conditions with green silicon carbide, which is used as the main
abrasive in the grinding of titanium alloys.

A comparison of the stress distributions plotted in Figs. 67–
69 shows that better results are obtained in respect of the magnitude and sign of the residual stresses if green silicon carbide
is replaced with zirconium carbide or tungsten carbide in the
grinding of titanium.

The fatigue strength of the samples of the alloy VTZ-1 ground
with green silicon carbide and tungsten boride is plotted in Fig.
70. We can see from this figure that when tungsten boride is used,
the fatigue limit is 14% higher than the limit of the samples ground
with standard tapes carrying green silicon carbide.

The test grinding of titanium alloys with disks made of tungsten boride, synthetic diamond, and silicon and zirconium carbides in Bakelite binder has shown that the surface finish which
can be achieved with the new abrasive materials and the efficiency
and wear of the disks made of this material are not inferior to
those made of silicon carbide or synthetic diamond. It is evident

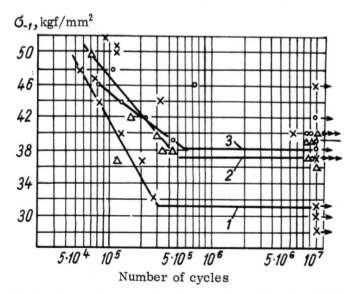

Fig. 70. Comparative curves showing the fatigue of the alloy VTZ-1
ground with different abrasives: 1) ZrC; 2) W_2B_5; 3) SiC.

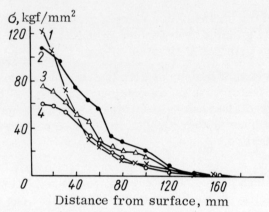

Fig. 71. Distribution of the stresses in the alloy VTZ-1
ground with different abrasives: 1) SiC; 2) ZrC; 3)
W_2B_5; 4) synthetic diamond.

from Fig. 71 that the residual macrostresses after grinding with
disks made of zirconium carbide and tungsten boride are 10-15
and 20-25% lower, respectively, than after grinding with disks
made of green silicon carbide.

The results of these experiments show that the use of new
abrasive materials will have less influence on the fatigue strength
of titanium alloys. It follows that under the extreme dry friction
conditions, such as high pressures and temperatures, the theoretical approach on the basis of the configurational model is quite
effective.

17.   Electronic Mechanism of Sintering

The current theoretical explanations of the sintering mechanism are based on diffusion, creep, and recrystallization in metals
and alloys [408-410], i.e., they are based on atomic phenomena
without allowance for the electronic structure. Nevertheless, it
is clear that diffusion, creep, and recrystallization as well as surface properties of substances are governed by the electron structure. Thus, the electron structure can provide a basis for a unified theory of all these phenomena and of the sintering mechanism
[403].

We may assume that the statistical weight of the stable configurations is always much lower on the surface of a sample than in
its bulk and the difference between the surface and bulk values in-

creases with increasing deformation of the bulk and distortion of
the surface. When two particles come into contact, thermal excitation gives rise to electron exchange which occurs in such a way
as to reestablish the disturbed stable configurations and to form
new such configurations at the interface between the particles.

When a single particle of irregular shape is heated, it tends
to assume a spherical shape so that the irregularities on the surface are removed. This occurs because of the tendency for reduction in the free energy, i.e., for increase in the statistical weight
of the stable configurations of the surface atoms.

Since metal particles are usually covered by oxide films, the
first stage of the sintering process is the chemical reduction of
these oxides which lowers the statistical weight of the stable configurations of the metal and increases correspondingly the fraction of the delocalized electrons. This process increases the activity of the surface atoms and intensifies the sintering process,
i.e., it intensifies the electron exchange between the metal atoms
on the surface of the same particle (this is usually called "surface
diffusion") and between the atoms belonging to different particles.
In both cases the statistical weight of the stable configurations increases. Obviously, this stage of the sintering process should be
intensified by a fall in the statistical weight of the stable configurations on the surfaces of the original particles and by a reduction
in the stability of these configurations. In the case of substances
which have the highest statistical weights of the stable configurations in the undisturbed state we find that deformation during
grinding and other methods of formation of particles and during
pressing has the least effect on the stable configurations. Consequently, we may expect the sintering process to be hardest to
achieve for these substances because this process involves reestablishment of the original stable configurations. If the statistical weight of the stable configurations of the atoms on the surfaces
of the original particles is high, the statistical weight of the intermediate configurations is low and this makes it difficult to
achieve electron exchange and sintering. The sintering processes
occur at relatively low temperatures, they are restricted to the
surfaces, and they eventually lead to an equilibrium state corresponding to the temperatures employed. Further sintering requires higher temperatures which disturb the stable configurations of the atoms and give rise to volume diffusion.

Considerations of the type outlined above explain why sintering does not occur or is difficult in the case of substances with the predominantly covalent directed type of bond such as graphite, boron nitride, silicon carbide, and diamond. Sintering of these materials requires high temperatures but at these temperatures the stable configurations, the directed bonds, and the nature of the substance are all destroyed because of disproportionation, decomposition, and transitions leading to structure modifications. If these processes are suppressed artificially (for example, by the use of high pressures) it becomes possible to sinter the substances just listed. Substances which are similar but have lower statistical weights of the stable configurations may be sintered in the usual manner: this applies to boron carbide which is characterized by a relatively low weight of the stable configurations, a moderate electrical resistivity, and a relatively narrow forbidden band. It is difficult to sinter pure oxides because of the high statistical weight of the stable configurations of the metal and oxygen atoms. The disturbance or excitation of these configurations is possible only at temperatures relatively close to the melting point.

It is interesting to note that the degree of thermal excitation required in the sintering of refractory carbides, borides, and silicides is governed primarily by the energy needed to disturb the sp^3 configurations which are responsible for the strongest $(Me-X)$ bonds in these materials. Therefore, it is found that carbides and borides can be sintered only at relatively high temperatures irrespective of the nature of the metal in these compounds. Silicides, in which the silicon has the sp^3 configuration, are less stable and, therefore, they can be sintered at low temperatures.

The activated sintering is always due to some chemical or physical excitation of a substance which disturbs the stable configurations, reduces their statistical weight, and then reestablishes them by the formation of a many-component system. Small admixtures of substances which can accept or donate delocalized electrons to the atoms of the substances being sintered can activate the sintering because they tend to increase the statistical weights of the stable configurations of all the components, i.e., to reduce the free energy of the whole system. The applicability of the configurational model to the interpretation of the changes in the properties which occur during the sintering of alloyed materials and which involve changes in the electron structure can be

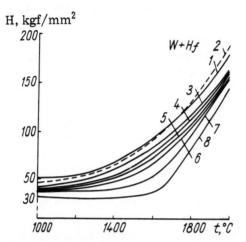

Fig. 72. Temperature dependences of the microhardness of tungsten alloyed with the following
amounts (wt.%) of Hf: 1) 0.05; 2) 0.1; 3) 0.2; 4)
0.3; 5) 0.4; 6) 0.5; 7) 1.0; 8) 2.0.

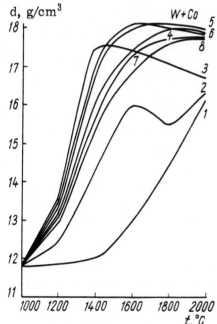

Fig. 73. Temperature dependences of the density of tungsten alloyed with the following
amounts (wt.%) of Co: 1) 0.05; 2) 0.1; 3) 0.2;
4) 0.3; 5) 0.4; 6) 0.5; 7) 1.0; 8) 2.0.

demonstrated by examining the recently obtained data on the influence of the alloying elements on the sintering temperature and the properties of tungsten [411-412]. Tungsten was alloyed in succession with almost all the elements in the three long periods and was sintered in the temperature range 1000-2000°C. The characteristics that can be conveniently compared are the microhardness (Fig. 72), the density (Fig. 73), and the strength. As expected, an increase in the sintering temperature in the 1000-2000°C range improves the strength and the density of the system finally obtained. The dependence of the density of alloyed ingots on the nature of the alloying element differs strongly for the group IV-V metals in the second half of the first long period beginning from chromium. The addition of Ti, Nb, Hf, and Ta reduces the density of tungsten after sintering whereas the addition of metals with $n_d \geq 5$ raises the density. It is clear that these relationships do not result from the differences between the experimental conditions (these were kept the same for all the alloying elements) but from the electron structure of the alloying atoms.

Since the statistical weight of the d^5 configurations in metallic tungsten is high (about 96%) and about one electron/atom is in the delocalized s state, whereas the atoms of the elements that reduce the density have d shells which are less than half-filled, it follows that introduction of such atoms into tungsten will result in the formation of the stable d^5 configurations of these atoms at the expense of their own s electrons and the delocalized electrons of tungsten. This will disturb the equilibrium in a crystal and reduce somewhat the statistical weight of the d^5 configurations of tungsten, in accordance with the general Hume-Rothery rules. Such changes in the electron structure should result in a fall of the strength of bonds (with increasing concentration of the alloying element) and in corresponding changes in the other properties.

The dependence of the density of the alloyed tungsten on the nature of the alloying element is also of interest. As expected, alloying with tantalum, which is characterized by a strong tendency to form the stable d^5 configurations, results in a stronger disturbance of the d^5 configurations of tungsten and, consequently, a greater decrease in the density and microhardness than that produced by alloying with hafnium. The same bond-weakening effect is observed when the principal quantum number of the valence electrons of the alloying element rises (the quantum number increases from titanium to hafnium); this is a consequence of an increase in the stability of the valence shell of the alloying element in a given group.

The similarity of the electron structures of tungsten and of the alloying elements gives rise to an additional relationship: when the concentration of the alloying element is increased to the extent that the atoms of this element begin to interact with each other and their electrons form a configuration spectrum, it is found that the stable configurations in the impurity sublattice are the d^0 and d^5 states, i.e., the same states as in the case of pure tungsten. Hence, it follows that it should be possible to vary monotonically the density, microhardness, and other properties of alloyed tungsten by varying the concentration of the alloying element within wide limits (Fig. 72).

The situation is somewhat different when tungsten is alloyed with elements located in the second half of the first long period (Fig. 73). In this case the valence shell of the impurity atoms is more than half-filled and when the impurity concentration is increased the impurity electrons should form a spectrum of the d^5 and d^{10} configurations, whereas the electron structure of tungsten is characterized by the coexistence of the d^0 and d^5 configurations. Hence, it is clear that the property—composition dependences of the alloys of tungsten with the elements in the second half of the first long period cannot be monotonic, as confirmed by the experimental results.

It is natural to take the largest change in the density of a material as being directly related to the electron structure of the atoms of the alloying admixture. This change will vary with the impurity (admixture) concentration and with the sintering temperature. The highest value of the density achieved by alloying tungsten and the corresponding optimal temperatures and concentrations are listed in Table 39. An analysis of the results of the values given in this table shows that the greatest difference between the electron structures of the impurity and the matrix (tungsten) atoms is observed for nickel, palladium, and platinum because the atoms of these elements have to give up the largest numbers of electrons to the collective state in the tungsten lattice. This disturbs the equilibrium in tungsten, and increases somewhat the delocalized fraction of the valence electrons as well as the sta-

TABLE 39. Density of Tungsten Alloyed with Various Elements under Optimal Conditions

Element	c_{opt}, %	T_{opt}, °C	ρ, g/cm³	Element	c_{opt}, %	T_{opt}, °C	ρ, g/cm³	Element	c_{opt}, %	T_{opt}, °C	ρ, g/cm³
Cr	2.0	1600	16.6	—	—	—	—	—	—	—	—
Fe	0.5	1600	17.95	Ru	1.0	1600	17.4	Os	3.0	2000	16.1
Co	0.4	1600	18.1	Rh	0.5	1600	17.8	Ir	1.0	2000	16.7
Ni	0.4	1600	18,4	Pd	0.4	1800	18.4	Pt	0.4	2000	17.4

tistical weight of the d^5 configurations. Consequently, the strength of the resultant substance should be higher than that of pure tungsten, which is in agreement with the experimental observations. A relatively smaller increase in the density and microhardness can be expected from doping with cobalt and the effects will be even weaker for doping with iron. We must also remember that we are considering the maximum density of a material which is attainable for any concentration of the impurity and any sintering temperature. Since nickel and its analogs tend to form the stable d^{10} states and since this group of metals is least able to give up electrons (we are comparing Cr, Fe, Co, and Ni), it follows that the optimal sintering temperature of the group VIIIC elements should be the highest, in agreement with the data given in Table 39 for all the group VIII elements. In the case of the group VIIIA elements the similarity of the atomic configurations to the d^5 configuration of tungsten should result in the smallest transfer of electrons from any one atom. Such a transfer needs the smallest energy, which increases on transition to group VIIIC because the configurations of Ni, Pd, and Pt tend to form the stable d^{10} states. Under these conditions, the optimal impurity concentrations, which increase toward the middle of the period, should be the lowest for Ni, Pd, and Pt. This is in basic agreement with the results given in Table 39 because very nearly the highest densities are obtained for a wide range of impurity concentrations. In comparing the densities we must keep to the values obtained for optimal sintering temperatures and optimal impurity concentrations. If we take the values obtained at a fixed sintering temperature of 1000°C, we find that in most cases the density increases when the alloying elements of the chromium group are replaced with nickel, palladium, or platinum. This is a consequence of the fact that it is more difficult to detach electrons when nickel is used than when iron or chromium are employed.

Similar conclusions follow also from an analysis of the alloying of tungsten with elements of the second and third long periods. Since the stability of the d^n configurations increases with increasing principal quantum number of the valence electrons, the alloying with the group VII and VIIIC elements of the first long period (for which the ease of transfer of the d electrons is greatest) reduces the optimal sintering temperatures (compared with the temperatures applicable to tungsten alloyed with their analogs in the second and third long periods) and produces a greater gain in the density (this should be contrasted with the opposite effect obtained for the group IV-V elements). The alloying with rhenium and osmium has very little effect on the properties of tungsten [412].

18. Structure of Phase Diagrams

The reasons for the formation of various binary systems have been analyzed on many occasions. The usual explanation is based on the crystal stereochemistry principle. Unfortunately, this principle does not satisfactorily account for the nature of the interaction in binary systems. For example, the difference between the atomic radii of metals in the same group of the fourth and fifth periods is greater than the corresponding difference for the fourth and sixth periods and the smallest difference is observed for metals of the fifth and sixth periods. If we assume that ideal solid solutions are formed in the systems with the most favorable steric factors, we find that the metals belonging to the fourth and sixth periods should form solid solutions more easily than the metals belonging to the fourth and fifth periods. In fact, the reverse is observed. Thus, it is not sufficient to consider phase (equilibrium) diagrams simply from the steric point of view.

The different types of interaction between the components in binary systems can be explained on the basis of the configurational model of the electron structure of atoms.

Table 40 lists the metallochemical properties of the transition metals belonging to groups III-VI. These properties are taken from [414]. We shall discuss the interaction between the group III transition metals and the group IV-VI transition metals by considering the binary yttrium systems [223, 415] whose phase diagrams are plotted in Fig. 74. We can see that the solubility of the transition metals in yttrium decreases with increasing statistical weight of the d^5 configurations of these metals. However, the weakest interaction with yttrium is not exhibited by tungsten (in spite of the fact that its statistical weight of the d^5 states is highest) but by tantalum.

In considering the yttrium systems we must bear in mind that yttrium undergoes that $\alpha \rightarrow \beta$ transition at 1495°C: in this transition the structure changes from the hcp to the bcc type. This transition from a modification with a lower to one with a

TABLE 40. Metallochemical Properties of Group III-VI
Transition Metals

IIIA	IVA	VA	VIA
Sc (18)	Ti (43)	V (63)	Cr (73)
1 — Y, Ti (β), Zr 2 — Ti, Ta 3 — 4 — 5 —	1—Sc (β), Hf (α, β), V (β), Zr (α, β), Nb (β), Ta (β), Cr (β), Mo (β) 2—Sc (α), Y, La, V (α), Nb (α), Ta (α), Cr (α), Mo (α), W 3 — Cr 4 — 5 —	1 — β-Ti, Nb, Ta, Cr, Mo, W 2 — Y, La, Zr, Hf 3 — Zr, Hf, Ta 4 — 5 —	1 — β-Ti, V, Mo, W 2 — Y, La, α-Ti, Cr, Nb, Ta 3 —Ti, Zr, Hf, Nb, Ta 4— 5—
Y (22)	Zr (53)	Nb (76)	Mo (88)
1 — Sc, La (β) 2 — α-La, Ti, Zr, Hf, V, Nb, Ta, Cr, Mo, W 3 — 4 — Ti, Zr, Hf, V, Nb, Mo, W 5 —	1 — Ti (α,β), Hf (α, β), Nb (β), Sc (α,β), Ta (β) 2 — Y, V, Nb (α), Ta, Cr, Mo, W 3—V, Cr, Mo, W 4 — 5 —	1 — β-Ti, (α,β)-Zr, β-Hf, V, Ta, Mo, W 2—Y,α-La, α-Zr, α-Hf, Cr 3 — Cr 4 — La 5 —	1 — β-Ti, V, Nb, Ta, Cr, W 2 — α-Ti, Zr, Hf 3 — Zr, Hf 4 — 5 —
La (23)	Hf (55)	Ta (81)	W (94)
1 — Y (β) 2 — Y (α), Ti, V, Nb, Cr 3 — 4 — 5 —	1 — Ti, Zr, Nb (β), Ta (β) 2 — Y, V, Mo, W 3 — V, Cr, Mo, W 4 — 5 —	1 — β-Ti, β-Zr, β-Hf, V, Nb, Mo 2 — Sc, α-Ti, α-Zr, α-Hf, Cr 3 — V, Cr 4 — 5 —	1 — V, Nb, Ta, Cr, Mo 2 — Y, Ti, Zr, Hf 3 — Zr, Hf 4 — 5 —

Note. 1) Continuous solid solutions; 2) restricted solid solutions; 3) compounds; 4) eutectic mixtures; 5) no interaction. The figures in parentheses are the statistical weights of the d^5 stable configurations.

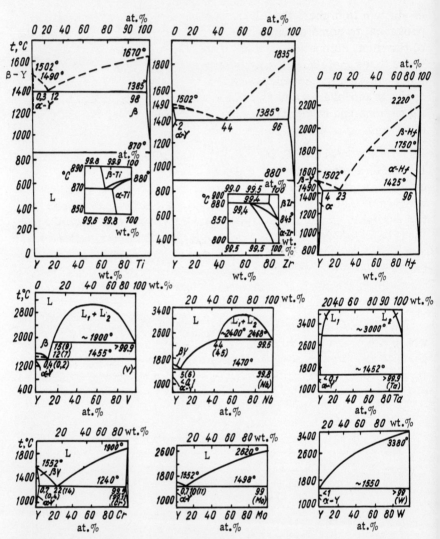

Fig. 74. Phase diagrams of systems formed by yttrium and group IV—VI transition metals.

higher symmetry is the result of an increase in the statistical weight of the d^5 configurations because of destruction of the less stable configurations.

Thus, yttrium participates either in the α or in the β form. Since the statistical weight of the d^5 configurations of the β form

of yttrium is higher, the electron structure of this form approaches, to some extent, the electron structures of the β phases
of titanium, zirconium, and hafnium. When yttrium interacts with
these three metals, it forms eutectic diagrams with extended
solid-solution (homogeneity) regions. These diagrams correspond to moderate values of the statistical weight of the d^5 configurations and to relatively high proportions of the delocalized
electrons.

Metals in group VI are usually acceptors capable of disturbing the d^5 configurations of the cores of the yttrium atoms. However, since these metals are characterized by high statistical
weights of their own stable configurations, they cannot accept all
the available yttrium electrons, some of which remain in the delocalized (collective) state. Consequently, binary eutectic systems
are formed with narrow solid-solution regions because the statistical weight of the d^5 configurations of these systems is higher
than the corresponding statistical weight of the systems formed
by yttrium with the group IV metals.

When yttrium interacts with the group V metals (V, Nb, Ta),
which are also acceptors and which disturb the stable configurations of the yttrium atoms, the statistical weight of the stable configurations of these metals may increase strongly at the expense
of the electrons provided by the disturbed yttrium configurations.

TABLE 41. Interaction of Yttrium with Group IV-VI
Transition Metals

Metal	Disturbance of d^5 con-figurations of yttrium atoms	Formation of d^5 con-figurations of partner atoms	Free-electron density	SWASC of d^5 states	Interaction between components
Ti	Weak	Weak	High	21	Limited solid solu-tions + eutectic
Zr	»	»	»	30	The same
Hf	»	»	»	33	» »
V	Moderate	Moderate	Moderate	41	Eutectic
Nb	»	»	»	54	»
Ta	Strong	Strong	Low	59	No interaction
Cr	Moderate to strong	Moderate to strong	Moderate to high	51	Limited solid solu-tions + eutectic
Mo	The same	Moderate to weak	»	66	The same
W	Strong	Weak	Moderate	72	» »

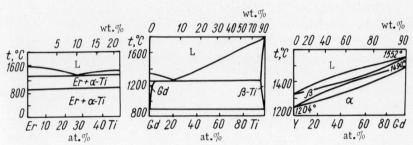

Fig. 75. Phase diagrams of systems formed by d- and f-type transition metals.

The degree of disturbance of the stable configurations of yttrium increases from vanadium to tantalum (Table 41) and this is accompanied by a very strong localization of the valence electrons. Consequently, the ratio of the components in the Y−Ta system may vary within wide limits and the solubility of the components in the solid state is low in all the systems in which the group V metals are involved.

Other rare-earth metals can act as donors or acceptors (Fig. 75). Erbium $(4f^{12}6s^2)$ has the f^{12} configuration in the free state but this is disturbed in the formation of metallic crystals in the direction of formation of the stable f^{14} configuration, i.e., erbium is an acceptor whereas titanium is a donor. In this system the density of the delocalized electrons is low and the solubility in the solid state is very limited.

Conversely, in the Gd−Ti system, in which Gd $(4f^75d^16s^2)$ is a donor of the 5d electrons, both components tend to acquire the stable f^7 configurations and act as donors so that extended solid-solution regions are observed. The enhancement of the donor properties on transition from titanium to yttrium results in the formation of a continuous series of solid solutions in the Gd−Y system. Alloys in this system are close to the ideal.

An examination of the phase diagrams of the $Me^{IV}− Me^{IV}$ systems (Fig. 76) shows that the nature of these diagrams is a function of the difference between the statistical weights of the stable configurations of the components: this difference is 9% for the Ti−Zr, 12% for the Ti−Hf and only 3% for the Zr−Hf systems. Consequently, the solubility in the last system is nearly ideal.

Similar conclusions can be drawn from an examination of the Me^V − Me^V and Me^{VI} − Me^{VI} systems. In the V−Ta system the solid solution precipitates below 1420°C forming the σ phase because of the strong donor tendency of vanadium and the strong acceptor tendency of tantalum. In the Cr−W system a complete mutual solubility is observed only at some temperatures and there is a wide region where the solubility is restricted. The difference between the values of the statistical weights of the stable configurations is greatest (18 and 21%) respectively, for the V−Ta and the Cr−W systems, whereas the corresponding difference for the other systems is within the range 5-15%. Consequently, the formation of continuous solid solutions is possible if the difference between the statistical weights of the stable configurations does not exceed 15%. When this difference becomes greater than 15%

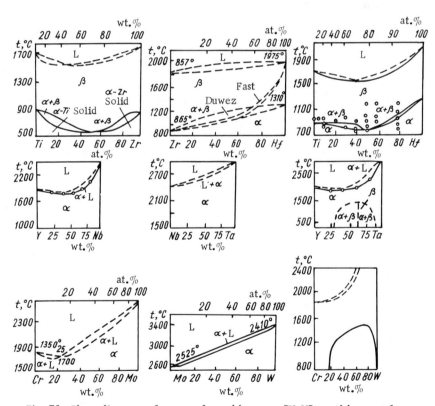

Fig. 76. Phase diagrams of systems formed by group IV-VI transition metals.

in the V—Ta and Cr—W systems, precipitation of the solid solution gives rise to the σ phase.

The limited solubility and the formation of chemical compounds in $Me^V - Me^{VI}$ systems are observed only for the Nb— Cr and Ta—Cr pairs. This is due to the acceptor properties of niobium and tantalum on the one hand, and the donor properties of chromium, on the other. The difference between the statistical weights of the unstable configurations for these two systems is lowest (3-8%) among the $Me^V - Me^{VI}$ systems. For the other systems of this type this difference is 10-31% (with the exception of the Ta—Mo system, for which it is 7%) and they form continuous solid solutions.

In the binary systems formed by group IV and V transition metals with the group VI metals a limited solubility in the solid state and the formation of chemical compounds are observed only for the V—Cr and V—Hf pairs (for these pairs the difference between the statistical weights of the stable configurations is 11 and 8%, respectively). The other systems of the same kind are characterized by differences between the statistical weights ranging from 20 to 38%. In systems formed by the group IV and VI metals the limited solid-solution regions correspond to differences between the statistical weights of the stable configurations in the range 36-51% and the formation of chemical compounds corresponds to differences ranging from 18 to 42%.

Similar conclusions are obtained for systems formed from titanium and the isomorphous elements V, Cr, Nb, Ta, and Mo. These systems differ in the extent of the corresponding phase regions, in the solubility of the elements in α titanium, and in the rate of fall of the $\beta \rightarrow \alpha$ transition temperature. The data on the solubility of these metals in α titanium at 600°C (Table 42) indicates that the solubility decreases with increasing statistical weight of the d^5 configurations.

When elements of groups V-VI are dissolved in titanium, the number of electrons that needs to be transferred increases from V to Cr, from Zr to Mo, and from Ta to W. The stability of the impurity atom configurations increases along the same direction and this hinders the transfer of electrons, which is manifested as a reduction in the solubility.

TABLE 42. Solubility of
Some Transition Metals
in α Titanium

Element	SWASC of d^5 states	Solubility in α Ti at 600°C, at.%
V	63	≈ 2.83
Nb	76	≈ 2
Ta	81	≈ 1.6
Mo	84	≈ 0.5
Cr	73	≈ 0.3

At certain concentrations of the second component in a system the conditions are optimal for the strongest localization of electrons in the stable configurations. The nature of these configurations is governed by the electron structure and by the electron exchange between the atoms of the components. The corresponding points in the phase diagrams can be calculated by a method suggested by Ul'yanov and Nazarenko [471].

In the transition-metal alloys the components can act as electron donors or acceptors, depending on the statistical weight of the stable configurations. The optimal ratio of the components for easy electron exchange is attained when the concentration of the acceptor in an alloy is proportional to the number of vacancies in the d shell which have to be filled to produce the stable d^5 configuration and the concentration of the donor is proportional to the number of the localized electrons:

$$\frac{X_A}{X_B} = \frac{5 - n_A}{n_B}. \tag{40}$$

The influence of the steric factor on the interaction between the components can be allowed for by introducing the ratio of the ionic radii for that degree of ionization which is equal to the number of the localized electrons of the components. Then, if we replace n_A with the statistical weight of the d^5 states (P_A), we obtain

$$\frac{X_A}{X_B} = \frac{100 - P_A}{P_B} \cdot \frac{r_B}{r_A}. \tag{41}$$

TABLE 43. Characteristic
Points in Phase Diagrams

System	Position of maximum, at.%		
	from phase diagram	calcu-lated	differ-ence
V — W	5.5 W	7.1 W	+1.6
Nb — W	5 W	7.6 W	+2.6
Cr — Mo	12 Mo	11.6 Mo	—0.4
Nb — Mo	20 Mo	15.4 Mo	—4.6
V — Nb	22 Nb	22.1 Nb	+0.1
Zr — Nb	22 Nb	32.3 Nb	+6.3
Zr — Ta	20 Ta	26 Ta	+6.0
V — Ta	17 Ta	17.4 Ta	+0.4
Hf — Ta	30 Ta	25.6 Ta	+0.4
Ti — V	28.7 V	48.0 V	+19.3
Cr — V	31 V	31.1 V	+0.1
Ti — Cr	45 Cr	43.5 Cr	—1.5
Hf — Ti	70 Hf	41.0 Hf	—29.0
Hf — Zr	35 Zr	41.8 Zr	6.8

The compositions of the alloys of refractory metals calculated by means of Eq. (41) correspond to the coordinate of
minima on the solidus-liquidus curves in the systems with complete mutual solubility of the components in one another and to
the coordinates of the eutectic points of the systems with limited
solubility. The calculated compositions of the alloys and the characteristic points in the phase diagrams are in good agreement
(Table 43). An analysis of the properties sensitive to the localized electron density in the alloys of refractory compounds
supports the validity of Eq. (41).

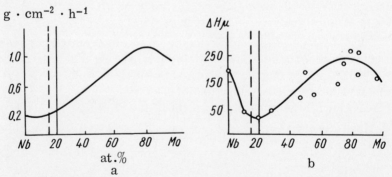

Fig. 77. Oxidation (a) and carbonization (b) curves of the Nb—Mo system.

A Zr−Nb alloy of the composition calculated by means of Eq. (41) has the lowest free energy and the highest heat of sublimation. An Nb−Mo alloy of similarity calculated composition has the minimum rate of oxidation and the minimum electrical resistivity. Minima in the oxidation rate curves are also observed for the Mo−W, V−Mo, Ta−Mo, and Ta−Nb alloys at the compositions which follow from Eq. (41). These conclusions were tested by investigating the high-temperature interaction in alloys of the Nb−W, Nb−V, and Nb−Mo systems with carbon. It was established that the interaction decreased on approach to the composition calculated from Eq. (41). Figure 77 shows the oxidation and carbonization curves of the Nb−Mo system. The alloy with the maximum localization of electrons, calculated from Eq. (41), is designated by a dashed vertical line. It is found that at this composition the interaction of the Nb−Mo system with carbon is weakest: carbide eutectics are not observed.

All these results show clearly that the strongest localization of electrons is observed in alloys whose compositions are predicted by Eq. (41) and this makes it possible to provide not only a qualitative but also a quantitative description of the features of phase diagrams of complex systems.

19. Catalytic Activity

Current theories of catalysis make it possible to select suitable catalysts on the basis of their solid-state properties. According to the electron theory the suitable properties are the type of conduction and the position of the Fermi level in semiconductors; the lattice parameter and the type of the lattice as well as the energies of bonds in the molecules reacting with a solid are used in the multiplet theory; the Lewis number of proton−donor centers, the e^2/r ratio of the cations, and their coordination number are used in the the theory of acid-base catalysis. However, none of these properties can be used as a universal criterion of the catalytic activity.

Transition metals are an important class of catalysts and it is worth investigating the role played by their electron structure in catalysis.

It is reported in [416, 417] that the catalytic activity of the transition metals with partly filled d and f electron shells is a consequence of their electron structure. A correlation between catalytic activity of the transition metals, their alloys, and compounds with the number of vacancies in the d shell of free atoms can be established on the basis of the donor—acceptor interaction in matter, which can be regarded as a consequence of the formation of stable configurations of the valence electrons.

The catalytic activity of the transition metals in any chemical process should be related in an unambiguous manner to the surface properties of the catalyst. It has often been pointed out [418, 419] that chemisorption plays the dominant role in heterogeneous catalysis. The result of such chemisorption is the liberation of free valences. The specific nature of the electron structure of the transition metals is manifested in the chemisorption stage.

We may assume that the catalytic action of the transition metals is governed by the stability of their partly filled shells because an increase in the catalytic activity corresponds to a reduction in the activation energy [420]. It is known that the activation energy is determined by the classical Coulomb energy E_C and by the exchange interaction energy E_{exc}. An increase in E_C reduces the activation barrier whereas an increase in E_{exc} raises this barrier [421]. Therefore, the catalytic activity of a transition metal can be raised by increasing E_C. This conclusion is well supported by the experimental results.

The configurational model can also explain the variations in the catalytic activity of the transition metals. This explanation can be provided if it is assumed that E_C is related directly to the polarizability of the bonds. Since the d^0, d^5, and d^{10} configurations are characterized by a spherical symmetry of the electron cloud, the polarization of these configurations is smallest. Therefore, the stable configurations of the localized fraction of the valence electrons are characterized by a low catalytic activity. This is supported by the observation that chromium, molybdenum, and tungsten are not good catalysts [421]. These three metals are characterized by the d^5 configurations with spherically symmetrical electron clouds. According to the Hund rules these metals should have five free valences which can give rise to directed covalent bonds. The weak polarizability of the bonds and the high numerical values of the statistical weights of the d^5 configurations are responsible for high activation energies and, consequently, low catalytic activity of chromium, molybdenum, and tungsten.

When the d shell contains $0 < n_d < 5$ electrons, the electron cloud is strongly aspherical and it is characterized by a very strong bond polarization, i.e., by a low activation energy E_a and a high catalytic activity. The rise of the catalytic activity from scandium to vanadium inclusive is pointed out in [422].

When n_d is close to $(2l + 1)$, where l is the orbital quantum number, the bond polarizability becomes weaker but this is compensated by the electrostatic deformation of the d^n electron configurations [423, 424] so that the resultant activation energy E_a decreases with increasing number of electrons right up to $n_d = 2l$. However, when $n_d = 2l + 1$ is reached, the spherical shape of the electron cloud reduces the catalytic activity.

Similar considerations apply also when the number of electrons in the d shell is $(2l + 1) < n_d < 2(2l + 1)$. In this case we again observe an increase in the activity from chromium to nickel inclusive and a sharp drop in the activity when copper is reached.

The catalytic activity of the d-type metals belonging to group VI is probably due to the van der Waals interaction and the electrostatic deformation of the electron cloud. In considering the van der Waals interaction in the transition metals with the highest values of the statistical weight of the stable electron configurations we can ignore, by analogy with the inert gases, the induction and orientation effects, and restrict the analysis to the quantum-mechanical (dispersion) interaction between the vibrating electrons [425].

When we go over to high values of the principal quantum number n the stability of the d^5 configurations increases but a considerable rise in the radius of the d orbit is accompanied by an increase in E_C, which reduces the activation energy and, consequently, increases the catalytic activity.

In considering the catalytic activity of alloys based on the transition metals, we can use the following classification:

    1)   alloys formed by combining the transition metals with similar proportions of their valence electrons localized in the stable configurations of the components;

2) alloys formed by combining the transition metals with strongly differing stabilities of their d configurations;
3) alloys of the transition metals with nontransition metals having the s valence electrons;
4) alloys of the transition metals with nontransition elements having the p valence electrons.

In the first case the statistical weight of the d^5 configurations cannot change significantly and, therefore the catalytic activity of such alloys obeys the same relationships as the activity of pure metals. In the case of alloys consisting of transition metals with strongly differing stabilities of the d configurations the conditions are favorable for the formation of stable states. Therefore, we may expect a reduction in the catalytic activity [426-428].

Alloys between the transition metals and nontransition metals having the s outer electrons are characterized by the formation (under favorable energy conditions) of the stable configurations of the localized valence electrons and, consequently, their catalytic activity should decrease. The conditions for the formation of the stable configurations should improve with decreasing ionization potential of the outer electrons, i.e., the activity should be lowest for the alloys formed between the transition metals and the alkali or alkaline-earth metals, copper or silver. For example, it is found that the catalytic activity of palladium−gold alloys in the ortho−para conversion of hydrogen decreases strongly when the concentration of gold exceeds 60% [427].

A similar fall in the activity is observed also when hydrogen is dissolved in the transition metals because this element can give up easily its s electrons to the d^0, d^5, and d^{10} stable configurations [429]. A reduction in the catalytic activity of nickel as a result of the addition of copper has been observed [430, 431] in investigations of the reactions involving hydrogen and, specifically, in the decomposition of methanol into CO and H_2 [432].

Alloys between the transition metals and nontransition elements having the p outer electrons are characterized by the formation of the stable sp^3 and s^2p^6 configurations. These configurations are formed at the expense of the delocalized valence electrons of the d-type metal: this alters the statistical weight of the d^n states and gives rise to the sp^3 configurations. A high catalytic activity can, therefore, be expected for the alloys formed between chromium and tungsten, which are not very active in the pure state. The reason for this is the disturbance of the stable d^n configurations in the formation of the more stable $s^x p^y$ configurations. This is confirmed by the strong catalytic activity of rhenium sulfides and selenides in the hydrogenation of unsaturated bonds in side branches of the derivatives of aromatic hydrocarbons. When we compare the catalytic activity of rhenium sulfides and selenides with that of molybdenum sulfide (Mo_2S_3), which is currently used in the hydrogenation of many classes of organic compounds, we find that rhenium compounds have definite advantages. In many cases the most active catalyssts are Re_2S_7 and Re_2Se_7, whereas ReS_2 and $ReSe_2$, though slightly inferior, are still tens of times more active than molybdenum sulfide [430, 431].

A high catalytic activity can thus be expected for aluminides, germanides, arsenides, tellurides, antimonides, and stannides of the metals with the highest values of the statistical weight of the d^5 stable configurations. The most important is the group of compounds of metals with typical nonmetals having the 2p and 3p valence electrons (boron, carbon, nitrogen, silicon, phosphorus, sulfur). The need to investigate these compounds (especially nitrides, carbides, and hydrides) was postulated on the basis of the multiplet theory in [433].

Boron and silicon have relatively low ionization potentials and, when they form compounds, they can attract the d electrons of the transition metals and establish configurations which are characterized by a low catalytic activity. A study of the properties of refractory compounds [434] in the dehydrogenation of cyclohexane by borides has established that these compounds are poor catalysts. The almost complete absence of the catalytic activity of NbB_2, ScB_2, YB_6, SrB_6, and HfB_6 has been established in [435]. Chromium borides are also completely inactive in the dehydrogenation of butylene into divinyl [433].

Alloys and compounds of phosphorus, carbon, and nitrogen with the transition metals exhibit a high catalytic activity. For example, titanium and zirconium carbides can catalyze the dehydrogenation of cyclohexane [433, 434]. Zirconium carbide is also a good catalyst in the hydrogenation of gaseous olefines [436].

A high activity can also be expected for many nitrides and phosphides. By way of example, we can point to the possibility of the dehydrogenation of butane to butylene by scandium nitride [433]. Alloys or compounds of this type are most likely to form the sp^3 and s^2p^6 configurations in the nonmetal sublattice and they can have a large contribution from the intermediate configurations. This leads to an enhancement of the catalytic activity provided the average d^n configuration is filled to the extent represented by $n_d \leq 2l$.

A similar explanation can be provided of the high activity of nickel boride (Ni_3B) in the hydrogenation reactions [429]; this explanation applies also to palladium, platinum, and ruthenium borides of the MeB_2 type. A high catalytic activity in the dehydrogenation of butylene into butenes and butadiene is exhibited by many chromium, molybdenum, and tungsten compounds such as carbides ($Cr_{23}C_6$, Cr_7C_3, Cr_3C_2), silicides (Cr_3Si, Cr_5Si_3, $CrSi$, $CrSi_2$) and by molybdenum and tungsten disilicides. Chromium nitrides (CrN and Cr_2N) and chromium phosphide (CrP) are also good catalysts but chromium bromides (CrB and CrB_2) are poor catalysts.

These results can be explained using the configurational model of solids. An increase in the concentration of boron in chromium boride increases the contribution of the covalent $B-B$ bonds to the total binding energy and improves the conditions for the formation of the d^5 configurations of chromium and the sp^3 configurations of boron. Chromium boride Cr_4B has a low concentration of boron and, therefore, a low catalytic activity, whereas the borides with higher concentration of boron (CrB and CrB_2) are completely inactive. On the other hand, the enhancement of the $Me-X$ bonds in the carbides, silicides, nitrides, and phosphides listed above is accompanied by the disturbance of the stable d^5 configurations of the metal atoms and the simultaneous increase in the statistical weight of the s^xp^y configurations, which increase the catalytic activity. Recently, the catalytic activity of compounds was studied in the model reaction of thermal decomposition of methane [437]. The catalysts were the carbides Cr_3C_2, Cr_7C_3, MoC, WC, and the nitride Cr_2N. The investigation included a study of the catalytic properties of iron and of finely dispersed nickel deposited on an aluminum oxide substrate.

The results obtained indicated that the lowest activity was exhibited by Mo_2C and WC, a higher activity was found for Cr_3C_2, Cr_7C_3, Cr_2N, and the highest activity was observed for nickel on Al_2O_3 and for iron.

The decomposition of methane consists of breaking of two-electron bonds between carbon and hydrogen. This breaking may occur in the presence of electron acceptors or substances which stimulate strong electron exchange, i.e., the materials which are characterized by a high statistical weight of the intermediate configurations. If the results just described are considered from this point of view, we find that molybdenum and tungsten carbides are characterized by the lowest weight of the d^n configurations of the intermediate spectra. This is why the catalytic activity of these two compounds is low.

The same reasoning accounts for the higher catalytic activity of the compounds of chromium, whose free atoms have the d^5s^1 configuration: the lower principal quantum number of this configuration implies that the stability of these configurations in chromium is lower than in analogs of this metal in the same group. Therefore, when chromium forms compounds the proportion of intermediate spectra may increase.

If we consider nickel, an atom of each has the d^8s^2 configuration in the free state, we may postulate that its high statistical weight of the intermediate configurations is also manifested in the high value of the electronic specific heat of this metal. The presence of configurations of the type found in nickel, which have the tendency to attract or give up electrons in order to form stable states, is responsible for the high catalytic activity. An even higher activity in the dissociation of methane is exhibited by iron which has a lower statistical weight of the intermediate configurations than that found in nickel. The highest activity of iron supports the hypothesis of the participation of iron carbide [438], which is undoubtedly formed in the catalytic process.

It is also worth mentioning other studies based on the configuration localization model [443, 444]: they are concerned with catalytic properties of rare-earth metals and their compounds (hydrides, carbides, and oxides) in the high-temperature para−ortho conversion of hydrogen. The role of the electron structure in the catalytic properties of metals in the hydrogenation reactions has been determined by Sokol'skii and Sokol'skaya [445]. A study of the influence of the admixtures of transition-metal carbides on the kinetics of chemical reduction of ferric oxide with hydrogen has established [446] that all carbides can catalyze this reduction. The catalytic activity of carbides increases with increasing atomic number of the metal component because of a reduction in the statistical weight of the stable sp^3 configurations and an increase of the statistical weight of the intermediate $s^x p^y$ configurations.

Literature Cited

1. E. M. Savitskii and V. V. Gribulya, Dokl. Akad. Nauk SSSR, 190:1147 (1970).
2. E. P. Gyftopoulos and G. N. Hatsopoulos, in: Direct Conversion of Thermal Energy into Electricity and Fuel Elements [Russian translation], VINITI, Moscow (1969), p. 134.
3. J. L. Beeby, Phys. Rev., 141:781 (1966).
4. A. L. Zilichikhis and Yu. P. Irkhin, Fiz. Tverd. Tela, 12:1981 (1970).
5. D. R. Penn, Phys. Rev., 142:350 (1966).
6. M. H. Cohen, in: Proc. Intern. School of Physics "Enrico Fermi," Course 37, Theory of Magnetism in Transition Metals, 1967, Academic Press, New York (1967), p. 403.
7. L. A. Maksimov and K. A. Kikoin, Fiz. Metal. Metalloved., 28:43 (1969).
8. J. C. Slater, Rev. Mod. Phys., 25:199 (1953).
9. E. C. Stoner, Proc. Roy. Soc. London, A154:656 (1936).
10. E. C. Stoner, Proc. Roy. Soc., A165:372 (1938).
11. E. C. Stoner, Proc. Roy. Soc., A169:339 (1939).
12. E. C. Stoner, Rep. Progr. Phys., 11:43 (1946-7).
13. E. C. Stoner, J. Phys. Radium, 12:372 (1951).
14. E. P. Wohlfarth, Rev. Mod. Phys., 25:211 (1953).
15. N. F. Mott, Proc. Phys. Soc. London, 47:571 (1935).
16. F. Seitz, Modern Theory of Solids, McGraw-Hill, New York (1940).
17. E. Wigner and F. Seitz, Phys. Rev., 43:804 (1933).
18. E. Wigner and F. Seitz, Phys. Rev., 46:509 (1934).
19. J. M. Ziman, Principles of the Theory of Solids, Cambridge University Press (1964).
20. J. Callaway, Energy Band Theory, Academic Press, New York (1964).
21. C. Herring, Phys. Rev., 57:1169 (1940).
22. N. F. Mott and H. Jones, The Theory of the Properties of Metals and Alloys, Clarendon Press, Oxford (1936); reprinted Dover, New York (1958).
23. D. H. Parkinson, Rep. Progr. Phys., 21:226 (1958).
24. M. F. Manning and M. I. Chodorow, Phys. Rev., 56:787 (1939).
25. N. F. Mott and K. W. H. Stevens, Phil. Mag., 2:1364 (1957).
26. C. Zener, Phys. Rev., 81:440 (1951).
27. K. Yosida, Phys. Rev., 106:893 (1957).

28.   C. Herring, J. Appl. Phys. Suppl., 31:3S (1960).
29.   W. Hume-Rothery, J. Inst. Metals, 35:295 (1926).
30.   H. Jones, Proc. Roy. Soc. London, A144:225 (1934); A147:396 (1934).
31.   P. W. Anderson, Phys. Rev., 124:41 (1961).
32.   W. Heisenberg, Z. Phys., 49:619 (1928).
33.   D. Pines and D. Bohm, Phys. Rev., 85:338 (1952).
34.   D. Bohm and D. Pines, Phys. Rev., 92:609 (1953).
35.   D. Pines and P. Nozières The Theory of Quantum Liquids, Vol. 1, Normal
      Fermi Liquids, Benjamin, New York (1966).
36.   G. V. Samsonov, Ukr. Khim. Zh., 31:1233 (1965).
37.   G. V. Samsonov, Porosh. Met., No. 12, p. 49 (1966).
38.   G. V. Samsonov, in: Sixth Plansee Seminar (High Temperature Materials),
      Reutte, Tyrol, 1968, publ. by Springer Verlag, Vienna and New York (1969),
      p. 10.
39.   G. V. Samsonov and I. F. Pryadko, Abstracts of Papers presented at All-Union
      Conf. on Solid State Electronics [in Russian], Kiev Polytechnic Institute (1967).
40.   Electron Structure of Transition Metals and Chemistry of Their Alloys [in
      Russian], Metallurgiya, Moscow (1966).
41.   S. V. Vonsovskii and Yu. A. Izyumov, Usp. Fiz. Nauk, 77:377 (1962); 78:3
      (1962).
42.   Yu. V. Mil'man, A. P. Rachek, and V. I. Trefilov, in: Problems in Physics
      of Metals and Metallography [in Russian], Naukova Dumka, Kiev (1965).
43.   R. J. Weiss and J. J. DeMarco, Rev. Mod. Phys., 30:59 (1958).
44.   B. W. Batterman, D. R. Chipman, and J. J. DeMarco, Phys. Rev., 122:68 (1961).
45.   M. I. Korsunskii and Ya. E. Genkin, Dokl. Akad. Nauk SSSR, 142:1276 (1962).
46.   M. I. Korsunskii and Ya. E. Genkin, Izv. Akad. Nauk SSSR, Neorg. Mater.,
      1:1701 (1965).
47.   M. I. Korsunskii, Ya. E. Genkin, and T. S. Verkhoglyadova, Porosh. Met.,
      No. 4, p. 35 (1962).
48.   M. I. Korsunskii and Ya. E. Genkin, Izv. Akad. Nauk SSSR, Ser. Fiz., 28:832 (1964).
49.   A. Z. Men'shikov and S. A. Nemnonov, Izv. Akad. Nauk SSSR, Ser. Fiz., 27:394
      (1963).
50.   G. Kemeny, Ann. Phys. (New York), 32:69 (1965).
51.   G. V. Samsonov, Yu. B. Paderno, and B. M. Rud', Izv. Vyssh. Ucheb. Zaved.,
      Fizika, No. 9, p. 129 (1967).
52.   J. C. Slater, Phys. Rev., 52:198 (1937).
53.   J. Friedel, G. Leman, and S. Olszewski, J. Appl. Phys. Suppl., 32:325S (1961).
54.   C. Herring and C. Kittel, Phys. Rev., 81:869 (1951).
55.   C. Herring, Phys. Rev., 85:1003 (1952).
56.   R. Kubo, T. Izuyama, D. J. Kim, and Y. Nagaoka, Proc. Intern. Conf. on Mag-
      netism and Crystallography, Kyoto, 1961, publ. in J. Phys. Soc. Japan,
      17(Suppl. B-I):67 (1962).
57.   J. L. Beeby, Proc. Phys. Soc. London, 90:779 (1967).
58.   W. M. Lomer and W. Marshall, Phil. Mag., 3:185 (1958).
59.   E. Wollan, Phys. Rev., 122:1710 (1961).
60.   J. B. Goodenough, Phys. Rev., 120:67 (1960).

61. J. Callaway and D. M. Edwards, Phys. Rev., 118:923 (1960).
62. R. E. Watson, Phys. Rev., 118:1036 (1960); 119:1934 (1960).
63. J. H. Van Vleck, Rev. Mod. Phys., 25:220 (1953).
64. J. Hubbard, Proc. Roy. Soc. London, A276:238 (1963).
65. D. E. Bondarev, Dokl. Akad. Nauk. SSSR, 154:83 (1964).
66. L. E. Orgel, Introduction to Transition-Metal Chemistry, Methuen, London (1960).
67. G. K. Boreskov, "Heterogeneous catalysis," in: Concise Chemical Encyclopaedia [in Russian], Vol. 2, Sovetskaya Éntsiklopediya, Moscow (1963), p. 468.
68. L. Pauling, Phys. Rev., 54:899 (1938).
69. L. Pauling, The Nature of the Chemical Bond and the Structure of Molecules and Crystals, 3rd ed., Cornell University Press, Ithaca, N. Y. (1960).
70. W. Hume-Rothery, H. M. Irving, and R. J. P. Williams, Proc. Roy. Soc. London, A208:431 (1951).
71. N. Engel, Ingeniøren N101 (1939); M1 (1940).
72. N. Engel, Haandbogi Metalläre, Selskabet for Metalforskning, Copenhagen (1945).
73. N. Engel, Kem. Maanedsbl., 30(5):53 (1949); 30(6):75 (1949); 30(8):97 (1949); 30(9):105 (1949); 30(10):114 (1949).
74. N. Engel, Powder Met. Bull., 7:8 (1954).
75. N. Engel, Amer. Soc. Metals Trans. Quart., 57:610 (1964).
76. N. Engel, Acta Met., 15:557 (1967).
77. N. Engel, in: Diffusion in Body-Centered Metals (Proc. Intern. Conf. Gatlinburg, Tenn., 1964), publ. by American Society for Metals, Metals Park, Ohio (1965), p. 87.
78. L. Brewer, in: Phase Stability in Metals and Alloys (Proc. First Batelle Memorial Institute Materials Science Colloquium, Geneva and Villars, Switzerland, 1966), McGraw-Hill, New York (1967), p. 39 (discussion pp. 241-249).
79. L. Brewer, ibid., p. 39 (discussion pp. 244, 560).
80. L. Brewer, High-Strength Materials (Proc. Second Berkeley International Materials Conference, University of California, Berkeley, 1964), Wiley, New York (1965), pp. 12-103.
81. L. Brewer, Acta Met., 15:553 (1967).
82. L. Brewer, in: Electron Structure of Transition Metals and Chemistry of Their Alloys [in Russian], Metallurgiya, Moscow (1966).
83. L. Brewer, Science, 161:115 (1968).
84. W. Hume-Rothery, Progr. Mater. Sci., 13:231 (1968).
85. W. G. Penney, The Quantum Theory of Valence, Methuen, London (1935).
86. R. S. Mulliken, Phys. Rev., 40:55 (1932).
87. I. F. Pryadko, Porosh. Met., No. 12, p. 64 (1966).
88. G. B. Dantzig, Econometrica, 17:200 (1949).
89. S. I. Zukhovitskii and L. I. Avdeeva, Linear Programming [in Russian], Nauka, Moscow (1964).
90. L. V. Kantorovich, Dokl. Akad. Nauk SSSR, 28:212 (1940).
91. G. V. Samsonov, I. F. Pryadko, B. M. Rud', V. V. Ogorodnikov, and L. F. Pryadko, Izv. Vyssh. Ucheb. Zaved., Fizika, No. 5, p. 120 (1970).

92. R. Nathans, C. G. Shull, G. Shirane, and A. J. Andersen, J. Phys. Chem.
     Solids, 10:138 (1959).
93. C. G. Shull and Y. Yamada, J. Phys. Soc. Japan, 17(Suppl. B-III):1 (1962).
94. R. Nathans and A. Paoletti, Phys. Rev. Lett., 2:254 (1959).
95. R. M. Moon, Phys. Rev., 136:A195 (1964).
96. I. Ya. Dekhtyar, et al., in: Investigations of Electronic Properties of Metals
     and Alloys [in Russian], Naukova Dumka, Kiev (1967).
97. G. V. Samsonov, Dokl. Akad. Nauk SSSR, 93:689 (1953).
98. V. Ern and A. C. Switendick, Phys. Rev., 137:A1927 (1965).
99. W. Hume-Rothery, Atomic Theory for Students of Metallurgy (Institute of
     Metals Monograph and Report Series, No. 3), Institute of Metals, London (1946).
100. F. Seitz, Phys. Rev., 47:400 (1935).
101. E. Gorin, Phys. Z. Sowjetunion, 9:328 (1936).
102. A. B. Neiding, Usp. Khim., 32:501 (1963).
103. A. B. Neiding, Priroda, No. 12, p. 111 (1964).
104. Compounds of Inert Gases [in Russian], Atomizdat, Moscow (1965).
105. V. S. Neshpor and G. V. Samsonov, Fiz. Metal. Metalloved., 25:1132 (1968).
106. G. V. Samsonov, I. F. Pryadko, and L. F. Pryadko, Izv. Vyssh. Ucheb. Zaved.,
     Fizika, No. 9, p. 60 (1969).
107. R. J. Weiss, Solid State Physics for Metallurgists, Addison-Wesley, Reading,
     Mass. (1963).
108. A. H. Sully, Chromium, Butterworths, London (1954).
109. G. E. Kimball, J. Chem. Phys., 8:188 (1940).
110. G. V. Samsonov, Beryllides [in Russian], Naukova Dumka, Kiev (1966).
111. V. S. Fomenko, Emission Properties of Elements [in Russian], Naukova Dumka,
     Kiev (1970).
112. L. D. Landau and E. M. Lifshitz, Quantum Mechanics: Non-Relativistic
     Theory, 2nd ed., Pergamon Press, Oxford (1965).
113. V. K. Grigorovich, Mendeleev's Periodic Law and Electron Structure of Metals
     [in Russian], Nauka, Moscow (1956).
114. I. R. Kozlova, V. N. Gurin, and A. P. Obukhov, Porosh. Met., No. 12, p. 68 (1966).
115. N. D. Borisov and V. V. Nemoshkalenko, Izv. Akad. Nauk SSSR, Ser. Fiz.,
     25:1002 (1961).
116. Yu. A. Izyumov and R. P. Ozerov, Magnetic Neutron Diffraction, Plenum
     Press, New York (1970).
117. J. H. Wood, Phys. Rev., 117:714 (1960).
118. F. Stern, Phys. Rev., 116:1399 (1959).
119. M. A. Blokhin, Physics of X Rays [in Russian], GITTL, Moscow (1957).
120. É. E. Vainshtein (Wainstein), Dokl. Akad. Nauk SSSR, 40:102 (1943).
121. W. W. Beeman and H. Friedman, Phys. Rev., 56:392 (1939).
122. V. V. Nemoshkalenko, Dokl. Akad. Nauk SSSR, 148:78 (1963).
123. A. Z. Men'shikov and S. A. Nemnonov, Fiz. Metal. Metalloved., 19:57 (1965).
124. V. V. Nemoshkalenko and V. Ya. Nagornyi, Izv. Akad. Nauk SSSR, Ser. Fiz.,
     31:990 (1967).
125. H. W. B. Skinner, T. G. Bullen, and J. E. Johnston, Phil. Mag., 45:1070 (1954).

126. M. A. Blokhin, V. F. Demekhin, and I. G. Shveitser, Izv. Akad. Nauk SSSR, Ser. Fiz., 28:834 (1964).
127. E. Noreland, Ark. Fys., 26:341 (1964).
128. N. D. Borisov and V. V. Nemoshkalenko, Izv. Akad. Nauk SSSR, Ser. Fiz., 25:1002 (1961).
129. J. E. Holliday, Phil. Mag., 6:801 (1961).
130. J. E. Holliday, J. Appl. Phys., 33:3259 (1962).
131. A. P. Lukirskii and T. M. Zimkina, Izv. Akad. Nauk SSSR, 27:330 (1963).
132. G. V. Samsonov, Yu. B. Paderno, and B. M. Rud', Izv. Vyssh. Ucheb. Zaved., Fizika, No. 9, p. 129 (1967).
133. M. A. Blokhin and I. Ya. Nikiforov, Izv. Akad. Nauk SSSR, Ser. Fiz., 28:780 (1964).
134. A. P. Lukirskii and I. A. Brytov, Izv. Akad. Nauk SSSR, Ser. Fiz., 28:841 (1964).
135. L. D. Landau and E. M. Lifshitz, Statistical Physics, 2nd ed., Pergamon Press, Oxford (1969).
136. F. F. Vol'kenshtein, Usp. Fiz. Nauk, 43:11 (1951).
137. W. A. Harrison, Pseudopotentials in the Theory of Metals, Benjamin, New York (1966).
138. J. C. Slater, Quantum Theory of Molecules and Solids, Vol. 1, Electronic Structure of Molecules, McGraw-Hill, New York (1963).
139. S. V. Vonsovskii, Zh. Eksp. Teor. Fiz., 16:981 (1946).
140. S. V. Vonsovskii and E. A. Turov, Zh. Eksp. Teor. Fiz., 24:419 (1953).
141. E. A. Turov and S. V. Vonsovskii, Zh. Eksp. Teor. Fiz., 24:501 (1953).
142. E. A. Turov and Yu. P. Irkhin, Fiz. Metal. Metalloved., 9:488 (1960).
143. Yu. P. Irkhin, Fiz. Metal. Metalloved., 6:214, 586 (1958).
144. E. A. Turov, Fiz. Metal. Metalloved., 6:203 (1958).
145. S. V. Vonsovskii, Modern Magnetism [in Russian], Gostekhizdat, Moscow (1953).
146. P. A. Wolff, Phys. Rev., 124:1030 (1961).
147. A. A. Abrikosov, I. P. Gor'kov, and M. E. Dzyaloshinskii, Methods of Quantum Field Theory in Statistical Physics, Prentice-Hall, Englewood Cliffs, N. J. (1963).
148. N. F. Mott, Proc. Phys. Soc. London, A62:416 (1949).
149. I. M. Lifshits and M. I. Kaganov, Usp. Fiz. Nauk, 69:419 (1959); 78:411 (1962).
150. N. M. Hugenholtz and L. van Hove, Physica (The Hague), 24:363 (1958).
151. M. Gell-Mann and K. A. Brückner, Phys. Rev., 106:364 (1957).
152. M. Gell-Mann, Phys. Rev., 106:369 (1957).
153. M. C. Gutzwiller, Phys. Rev., 137:A1726 (1965).
154. P. Lederer and A. Blandin, Phil. Mag., 14:363 (1966).
155. V. Heine, in: Low Temperature Physics — LT9 (Proc. Ninth Intern. Conf., Columbus, Ohio, 1964), Part B, Plenum Press, New York (1965), p. 698.
156. A. A. Sokolov, Quantum Mechanics [in Russian], Prosveshchenie, Moscow (1965).
157. C. Kittel, Elementary Solid State Physics, Wiley, New York (1962).

158.   L. S. Darken and R. W. Gurry, Physical Chemistry of Metals, McGraw-Hill, New York (1953).

159.   E. Dempsey, Phil. Mag., 8:285 (1963).

160.   B. T. Matthias, E. A. Wood, E. Corenzwit, and V. B. Bala, J. Phys. Chem. Solids, 1:188 (1956).

161.   I. I. Kornilov, Metalloids and Interactions between Them [in Russian], Nauka, Moscow (1964), p. 86.

162.   D. M. Yost, H. Russell, and C. Garner, The Rare-Earth Elements and Their Compounds, Chapman and Hall, London; Wiley, New York (1947).

163.   G. S. Zhdanov, Crystal Physics, Oliver and Boyd, Edinburgh (1965).

164.   I. F. Pryadko, Zh. Fiz. Khim., 42:755 (1968).

165.   I. F. Pryadko, Abstracts of Papers presented at Fifth Inter-Institute Seminar on Physical Properties and Electron Structure of Transition Metals, Their Compounds, and Alloys [in Russian], Kiev Polytechnic Institute (1966), p. 7.

166.   E. U. Condon and G. H. Shortley, The Theory of Atomic Spectra, Cambridge University Press; MacMillan, New York (1935).

167.   H. A. Bethe, Intermediate Quantum Mechanics, Benjamin, New York (1964).

168.   Ya. S. Umanskii, Physical Metallography [in Russian], Prosveshchenie, Moscow (1955).

169.   C. A. Coulson, Valence, Oxford University Press, London (1961).

170.   M. G. Veselov, Elementary Quantum Theory of Atoms and Molecules [in Russian], Fizmatgiz, Moscow (1962).

171.   J. B. Goodenough, Magnetism and the Chemical Bond, Interscience, New York (1963).

172.   M. I. Korsunskii and Ya. E. Genkin, in: Electron Structure of Transition Metals [in Russian], Scientific-Technical Information Division, Institute of Metal Physics, Academy of Sciences of the Ukrainian SSR, Kiev (1968).

173.   W. Hume-Rothery, Progr. Mater. Sci., 13:231 (1967).

174.   G. V. Samsonov, Ukr. Khim. Zh., 33:763 (1967).

175.   A. J. Cornish, Nature, 211:512 (1966).

176.   J. C. Slater, Quantum Theory of Molecules and Solids, Vol. 3, Insulators, Semiconductors, and Metals, McGraw-Hill, New York (1967).

177.   G. V. Samsonov (ed.), Physicochemical Properties of Elements [in Russian], Naukova Dumka, Kiev (1965). [Handbook of the Physicochemical Properties of the Elements, IFI/Plenum, New York (1968).]

178.   M. A. Filyand and E. I. Semenova, Properties of Rare-Earth Elements [in Russian], Metallurgiya, Moscow (1964).

179.   G. V. Samsonov, Dokl. Akad. Nauk SSSR, 180:377 (1968).

180.   G. V. Samsonov, Refractory Compounds of Rare-Earth Metals with Nonmetals [in Russian], Metallurgiya, Moscow (1964).

181.   G. V. Samsonov and V. Ya. Shlyukov, Ukr. Fiz. Zh., 11:437 (1966).

182.   C. A. Swenson, "Physics at high pressure," Solid State Phys., 11:41 (1960).

183.   P. M. Ogibalov and I. A. Kiiko, Behavior of Matter under Pressure [in Russian], Moscow State University (1962).

184.   G. V. Samsonov and V. I. Shesternenkov, Vestn. Kiev. Politekh. Inst., Ser. Mashinostr., No. 6, p. 152 (1969).

185.   L. E. Orgel, Introduction to Transition-Metal Chemistry, Methuen, London (1960).
186.   S. M. Braun, Abstracts of Papers presented at Fourth Conf. of Postgraduate Students and Young Scientists [in Russian], Scientific-Technical Information Division, Institute of Problems in Materials Science, Academy of Sciences of the Ukrainian SSR, Kiev (1970).
187.   M. J. McQuillan, "Phase transformations in titanium and its alloys," Met. Rev., 8:41-104 (1963).
188.   Ya. S. Malakhov and I. G. Tkachenko, Abstracts of Papers presented at Seventh Symposium on Physical Properties and Electron Structure of Transition Metals and Their Alloys and Compounds [in Russian], Scientific-Technical Information Division, Institute of Problems in Materials Science, Academy of Sciences of the Ukrainian SSR, Kiev (1969).
189.   G. V. Samsonov, Ukr. Khim. Zh., 31:433 (1965).
190.   G. V. Samsonov, Beryllides [in Russian], Naukova Dumka, Kiev (1966).
191.   G. V. Samsonov, et al., Boron, Its Compounds and Alloys [in Russian], Academy of Sciences of the Ukrainian SSR, Kiev (1960).
192.   G. V. Samsonov, Ukr. Khim. Zh., 31:1005 (1965).
193.   G. V. Samsonov, Nitrides [in Russian], Naukova Dumka, Kiev (1968).
194.   G. V. Samsonov, in: Chalcogenides [in Russian], Naukova Dumka, Kiev (1967).
195.   V. P. Perminov and G. V. Samsonov, Porosh. Met., No. 6, p. 41 (1966).
196.   V. S. Sinel'nikova and A. S. Goral'nik, Porosh. Met., No. 10, p. 51 (1968).
197.   G. V. Samsonov, Silicides [in Russian], Metallurgizdat, Moscow (1964).
198.   V. N. Bondarev and G. V. Samsonov, Porosh. Met., No. 6, p. 52 (1966).
199.   M. P. Arbuzov and B. V. Khaenko, Porosh. Met., No. 4, p. 74 (1969).
200.   L. Ramqvist, K. Hamrin, G. Johansson, A. Fahlman, and C. Nordling, J. Phys. Chem. Solids, 30:1835 (1969).
201.   L. Ramqvist, Jernkontorets Ann., 153:159 (1969).
202.   L. Ramqvist, Jernkontorets Ann., 152:465 (1968).
203.   Ya. S. Malakhov and I. G. Tkachenko, Porosh. Met., No. 1, p. 58 (1967).
204.   H. Montgomery and G. P. Pells, Proc. Phys. Soc. London, 78:622 (1961).
205.   L. M. Roberts, in: Low Temperature Physics and Chemistry (Proc. Fifth Intern. Conf., University of Wisconsin, Madison, 1957), publ. by University of Wisconsin Press, Madison (1958), p. 47.
206.   B. J. C. van der Hoeven and P. H. Keesom, Bull. Amer. Phys. Soc., 9:268 (1964).
207.   T. R. Waite, R. S. Craig, and W. E. Wallace, Phys. Rev., 104:1240 (1956).
208.   K. Huang, Statistical Mechanics, Wiley, New York (1963).
209.   N. M. Wolcott, Phil. Mag., 2:1246 (1957).
210.   F. E. Hoare and B. Yates, Proc. Roy. Soc. London, A240:42 (1957).
211.   H. Mongomery and G. P. Pells, Proc. Phys. Soc. London, 78:622 (1961).
212.   A. Berman, M. W. Zemansky, and H. A. Boorse, Phys. Rev., 109:70 (1958).
213.   O. V. Lounasmaa, Phys. Rev., 126:1352 (1962).
214.   O. V. Lounasmaa and P. R. Roach, Phys. Rev., 128:622 (1962).
215.   O. V. Lounasmaa and R. A. Guenther, Phys. Rev., 126:1357 (1962).

216.   O. V. Lounasmaa, Phys. Rev., 133:A219 (1964).
217.   O. V. Lounasmaa, Phys. Rev., 133:A502 (1964); 134:AB1 (1964).
218.   O. V. Lounasmaa, Phys. Rev., 133:A211 (1964).
219.   O. V. Lounasmaa, Phys. Rev., 134:A1620 (1964).
220.   V. V. Fesenko, Abstract of Doctoral Thesis [in Russian], Institute of Problems
       in Materials Science, Academy of Sciences of the Ukrainian SSR, Kiev (1968).
221.   A. S. Bolgar, Abstract of Thesis for Candidate's Degree [in Russian], Institute
       of Problems in Materials Science, Academy of Sciences of the Ukrainian SSR,
       Kiev (1967).
222.   S. P. Gordienko, Abstract of Thesis for Candidate's Degree [in Russian], Institute
       of Problems in Materials Sciences, Academy of Sciences of the Ukrainian SSR,
       Kiev (1967).
223.   E. M. Savitskii and G. S. Burkhanov, Metallography of Refractory Alloys and
       Rare-Earth Metals [in Russian], Nauka, Moscow (1971).
224.   J. J. Gilman, in: The Physics and Chemistry of Ceramics (Proc. Symp.,
       Pennsylvania State University, 1962), Gordon and Breach, New York (1963),
       p. 240.
225.   G. V. Samsonov, Usp. Khim., 35:779 (1966).
226.   É. E. Vainshtein, S. M. Blokhin, and P. I. Kripyakevich, Dokl. Akad. Nauk
       SSSR, 142:85 (1962).
227.   A. Zalkin, R. G. Bedford, and D. E. Sands, Acta Crystallogr., 12:700 (1959).
228.   T. Ya. Kosolapova, Carbides [in Russian], Metallurgiya, Moscow (1968).
229.   S. A. Nemnonov and A. Z. Men'shikov, Izv. Akad. Nauk SSSR, Ser. Fiz.,
       23:578 (1959).
230.   Ya. S. Malakhov and G. V. Samsonov, Porosh. Met., No. 12, p. 84 (1966).
231.   G. N. Dubrovskaya, Abstract of Thesis for Candidate's Degree [in Russian],
       Institute of Problems in Materials Science, Academy of Sciences of the
       Ukrainian SSR, Kiev (1967).
232.   R. B. Kotel'nikov et al., Highly Refractory Compounds [in Russian], Metal-
       lurgiya, Moscow (1969).
233.   A. Ya. Kuchma and G. V. Samsonov, Izv. Akad. Nauk SSSR, Neorg. Mater.,
       2:1970 (1966).
234.   G. V. Samsonov and V. G. Grebenkina, Porosh. Met., No. 2, p. 35 (1968).
235.   V. G. Grebenkina and A. Ya. Kuchma, Izv. Akad. Nauk SSSR, Neorg. Mater.,
       5:717 (1969).
236.   V. G. Grebenkina and E. N. Denbnovetskaya, Porosh. Met., No. 3, p. 34
       (1968).
237.   A. Ya. Kuchma, Izv. Akad. Nauk SSSR, Neorg. Mater., 3:884 (1967).
238.   G. V. Samsonov and A. Ya. Kuchma, Izv. Akad. Nauk SSSR, Neorg. Mater.,
       4:1361 (1968).
239.   G. V. Samsonov, Abstracts of Papers presented at Seventh Symposium on
       Physical Properties and Electron Structure of Transition Metals, Their Alloys,
       and Compounds [in Russian], Scientific-Technical Information Division,
       Institute of Problems in Materials Science, Academy of Sciences of the
       Ukrainian SSR, Kiev (1969).
240.   L. Ramqvist, Jernkontorets Ann., 152:465, 517 (1968).

241. G. V. Samsonov and G. Sh. Upadkhaya, in: Sixth Plansee Seminar (High-Temperature Materials), Reutte, Tyrol, 1968, Springer Verlag, Vienna and New York (1969).

242. G. V. Samsonov and G. Sh. Upadkhaya, Porosh. Met., No. 5, p. 69 (1969).

243. A. L. Giorgi, E. G. Szklarz, E. K. Storms, A. L. Bowman, and B. T. Matthias, Phys. Rev., 125:837 (1962).

244. S. S. Ordan'yan, A. I. Avgustinik, and L. V. Kudryasheva, Porosh. Met., No. 8, p. 26 (1968).

245. C. P. Kempter and E. K. Storms, J. Less-Common Metals, 13:443 (1967).

246. T. A. Sandenaw and E.K. Storms, J. Phys. Chem. Solids, 27:217 (1966).

247. G. V. Samsonov, A. Ya. Kuchma, and I. I. Timofeeva, Porosh. Met., No. 9, p. 69 (1970).

248. L. A. Reznitskii, Izv. Akad. Nauk SSSR, Neorg. Mater., 2:953 (1966).

249. V. Ya. Naumenko, Abstracts of Papers presented at Fourth Postgraduate Students Conference [in Russian], Institute of Problems in Materials Science, Academy of Sciences of the Ukrainian SSR, Kiev (1970).

250. É. Z. Kurmaev, S. A. Nemnonov, A. Z. Men'shikov, and G. P. Shveikin, Izv. Akad. Nauk SSSR, Ser. Fiz., 31:996 (1967).

251. E. N. Denbnovetskaya, Abstract of Thesis for Candidate's Degree [in Russian], Institute of Problems in Materials Science, Academy of Sciences of the Ukrainian SSR, Kiev (1967).

252. H. Bilz, Z. Phys., 153:338 (1958).

253. V. N. Bondarev, Abstract of Thesis for Candidate's Degree [in Russian], Institute of Problems in Materials Science, Academy of Sciences of the Ukrainian SSR, Kiev (1968).

254. G. V. Samsonov, M. I. Murguzov, and Yu. B. Paderno, in: Proc. All-Union Symposium on Physical Properties and Electron Structure of Rare-Earth Metals [in Russian], Scientific-Technical Information Division, Institute of Problems in Materials Science, Academy of Sciences of the Ukrainian SSR, Kiev (1968).

255. V. G. Grebenkina, Abstract of Thesis for Candidate's Degree [in Russian], Institute of Problems in Materials Science, Academy of Sciences of the Ukrainian SSR, Kiev (1969).

256. G. V. Samsonov, Yu. B. Paderno, and V. S. Fomenko, Zh. Tekh. Fiz., 36:1435 (1966).

257. F. Rother and H. Bomke, Z. Phys., 86:231 (1933).

258. F. Rother and H. Bomke, Z. Phys., 87:806 (1934).

259. E. H. B. Bartelink, Physica (The Hague), 3:193 (1936).

260. H. Bomke, Z. Phys., 90:542 (1934).

261. J. F. Chittum, J. Phys. Chem., 38:79 (1934).

262. W. M. H. Sachtler, Z. Elektrochem., 59:119 (1955).

263. V. V. Demchenko and N. E. Khomutov, Tr. Mosk. Inzh.-Fiz., 39:115 (1962).

264. P. Gombas, Nature, 157:668 (1946).

265. S. N. Zadumkin, Zh. Fiz. Khim., 27:502 (1953).

266. L. L. Kunin, Dokl. Akad. Nauk SSSR, 79:93 (1951).

267. R. M. Vesenin, Zh. Fiz. Khim., 27:878 (1953).

268. I. A. Podchernyaeva, G. V. Samsonov, and V. S. Fomenko, Izv. Vyssh. Ucheb.
      Zaved., Fizika, No. 6, p. 42 (1969).
269. G. V. Samsonov, V. S. Neshpor, and G. A. Kudintseva, Radiotekh. Elektron.,
      2:631 (1957).
270. G. V. Samsonov, V. S. Fomenko, and Yu. A. Kunitskii, Radiotekh. Elektron.,
      16:1304 (1971).
271. N. I. Siman, Abstracts of Papers presented at Fourth Conf. of Postgraduate
      Students and Young Scientists [in Russian], Scientific-Technical Information
      Division, Institute of Problems in Materials Science, Academy of Sciences
      of the Ukrainian SSR, Kiev (1970).
272. G. V. Samsonov, S. P. Gordienko, and I. A. Podchernyaeva, Elektron. Obrab.
      Mater., No. 1, p. 56 (1968).
273. I. A. Podchernyaeva, G. V. Samsonov, and V. S. Fomenko, Dokl. Akad.
      Nauk Ukr. SSR A, No. 4, p. 369 (1967).
274. I. A. Podchernyaeva, Abstract of Thesis for Candidate's Degree [in Russian],
      Institute of Problems in Materials Science, Academy of Sciences of the
      Ukrainian SSR, Kiev (1967).
275. G. V. Samsonov, I. M. Mukha, and A. N. Krushinskii, Elektron. Obrab. Mater.,
      No. 1, p. 28 (1966).
276. A. D. Verkhoturov, Elektron. Obrab. Mater., No. 1, p. 25 (1969).
277. G. V. Samsonov and I. M. Mukha, Porosh. Met., No. 7, p. 81; No. 8, p. 84
      (1968).
278. L. S. Palatnik, Dokl. Akad. Nauk SSSR, 89:455 (1953).
279. V. Yu. Veroman, Wear Resistance of Tools Used in Electro-Erosion Machining
      [in Russian], LDNTP, Leningrad (1964).
280. K. Albinski, Stanki Instrum., 35(7):11 (1964).
281. I. A. Kaptsov, Electronics [in Russian], GITTL, Moscow (1954).
282. P. G. de Gennes, Superconductivity of Metals and Alloys, Benjamin, New York
      (1966).
283. T. K. Kononenko, G. V. Samsonov, and L. K. Shvedova, in: Abstracts of
      Papers presented at Seventh Symposium on Physical Properties and Electron
      Structure of Transition Metals, Their Alloys, and Compounds [in Russian],
      Scientific-Technical Information Division, Institute of Problems in Materials
      Science, Academy of Sciences of the Ukrainian SSR, Kiev (1969).
284. G. V. Samsonov, in: Metal Physics and Metallography of Superconductors
      [in Russian], Nauka, Moscow (1965).
285. G. V. Samsonov, Yu. K. Lapshov, I. A. Podchernyaeva, V. S. Fomenko,
      Yu. I. Erosov, and E. M. Dudnik, Izv. Akad. Nauk SSSR, Neorg. Mater., 2:1454
      (1966).
286. G. V. Samsonov and T. S. Verkhoglyadova, Dokl. Akad. Nauk SSSR, 138:342
      (1961).
287. J. Piper, in: Proc. Tenth Nuclear Metallurgy Symposium, University of
      Colorado, Boulder, 1964, Metallurgical Society of the American Institute of
      Metallurgical Engineers Division, IMD Special Report No. 13 (1964), p. 29.
288. Ya. G. Dorfman, Zh. Eksp. Teor. Fiz., 35:533 (1958).

289. E. Lieb and D. Mattis, J. Math. Phys. (New York), 3:749 (1962).
290. Some Problems in Electron Structure of Transition Metals [in Russian], Nau-
     kova Dumka, Kiev (1968).
291. I. E. Dzyaloshinskii, Zh. Eksp. Teor. Fiz., 47:336 (1964).
292. S. V. Vonsovskii and K. B. Vlasov, Zh. Eksp. Teor. Fiz., 25:327 (1953).
293. E. Wollan, in: Theory of Ferromagnetism of Metals and Alloys [in Russian],
     IL, Moscow (1963).
294. P. A. Rebinder, "Hardness," in: Technical Encyclopedia [in Russian], Vol. 22
     (1933), p. 703.
295. P. A. Rebinder et al., in: Investigations of Surface Phenomena [in Russian],
     ONTI, Moscow (1936).
296. V. D. Kuznetsov, Surface Energy of Solids [in Russian], GITTL, Moscow (1954).
297. G. V. Samsonov, A. A. Ivan'ko, and E. N. Chupakhina, Fiz.-Khim. Mekh.
     Mater., 2:152 (1966).
298. V. V. Stasovskaya, Abstract of Thesis for Candidate's Degree [in Russian],
     Institute of Problems in Materials Science, Academy of Sciences of the Uk-
     rainian SSR, Kiev (1967).
299. A. A. Ivan'ko, Hardness [in Russian], Naukova Dumka, Kiev (1968).
300. V. S. Neshpor and G. V. Samsonov, Fiz. Metal. Metalloved., 4:181 (1957).
301. C. Kittel, Introduction to Solid State Physics, 2nd ed., Wiley, New York
     (1956).
302. J. Friedel, Dislocations, Pergamon Press, Oxford (1964).
303. A. Seeger, in: Dislocations and Mechanical Properties of Crystals (Proc.
     Intern. Conf., Lake Placid, N. Y., 1956), Wiley, New York (1957).
304. V. E. Panin and V. P. Fadin, Izv. Vyssh. Ucheb. Zaved., Fizika, No. 3, p. 72
     (1968).
305. J. Spreadborough, Phil. Mag., 3:1167 (1958).
306. S. L. Altmann, C. A. Coulson, and W. Hume-Rothery, Proc. Roy. Soc.
     London, A240:145 (1957).
307. G. Sh. Upadkhaya, Izv. Vyssh. Ucheb. Zaved., Fizika, No. 8, p. 38 (1969).
308. R. L. Segall, Acta Met., 9:975 (1961).
309. A. S. Keh and S. Weissmann, in: Electron Microscopy and Strength of Crys-
     tals (Proc. First Berkeley Intern. Materials Conf., University of California,
     Berkeley, 1961), publ. by Interscience, New York (1963), p. 231.
310. J. S. Hirschhorn, J. Less-Common Metals, 5:493 (1963).
311. N. I. Noskova, V. A. Pavlov, and S. A. Nemnonov, Fiz. Metal. Metalloved.,
     20:920 (1965).
312. N. I. Noskova, S. A. Nemnonov, and V. A. Pavlov, in: Properties and Applica-
     tions of Refractory Alloys [in Russian], Nauka, Moscow (1966).
313. A. Howie and P. R. Swann, Phil. Mag., 6:1215 (1961).
314. B. Henderson, J. Inst. Metals, 92:55 (1963).
315. K. Nakajima, J. Phys. Soc. Japan, 14:1825 (1959).
316. K. Nakajima and K. Numakura, Phil. Mag., 12:361 (1965).
317. C. S. Barrett and T. B. Massalski, in: The Mechanism of Phase Transforma-
     tions (Proc. Symposium, London, 1955), Institute of Metals Monograph and
     Report Series No. 18, Institute of Metals, London (1956), p. 331 (discussion).

318. Ya. D. Vishnyakov, D. M. Mazo, and Ya. S. Umanskii, Izv. Vyssh. Ucheb. Zaved., Chern. Met., No. 9, p. 145 (1963).

319. Ya. D. Vishnyakov, G. Sh. Gasanov, and Ya. S. Umanskii, Izv. Akad. Nauk Azerb. SSR, Ser. Fiz.-Tekh. Mat. Nauk, No. 2, p. 108 (1967).

320. R. P. Stratton and W. J. Kitchingman, Brit. J. Appl. Phys., 16:1311 (1965).

321. D. V. Lotsko and V. I. Trefilov, Fiz. Metal. Metalloved., 19:891 (1965).

322. P. Coulomb, N. Gibert, A. Clement, and A. Coujou, J. Phys. (Paris), 27:94 (1966).

323. B. E. P. Beeston and L. K. France, J. Inst. Metals, 96:105 (1968).

324. R. W. K. Honeycombe, J. S. T. van Answegen, D. H. Warrington, in: Structure and Mechanical Properties of Metals [Russian translation], Metallurgiya, Moscow (1967), p. 172.

325. G. V. Samsonov and G. Sh. Upadkhaya, Porosh. Met., No. 9, p. 70 (1968).

326. J. L. Martin, B. Jouffrey, and P. Costa, Phys. Status Solid, 22:349 (1967).

327. K. A. Osipov, Some Activated Processes in Solid Metals and Alloys [in Russian], Izd. AN SSSR, Moscow (1962).

328. G. V. Samsonov and S. A. Bozhko, Porosh. Met., No. 3, p. 35 (1970).

329. E. M. Savitskii, Rare-Earth Metals and Alloys [in Russian], Izd. Doma Tekhniki, Moscow (1959).

330. E. M. Savitskii and M. A. Tylkina, in: Investigations of Refractory Alloys [in Russian], Vol. 4, Izd. AN SSSR, Moscow (1959), p. 218.

331. A. A. Bochvar, Metallography [in Russian], Metallurgizdat, Moscow (1956).

332. I. P. Kushtalova and A. N. Ivanov, Porosh. Met., No. 9, p. 81 (1966).

333. I. P. Kushtalova and A. N. Ivanov, Porosh. Met., No. 2, p. 13 (1967).

334. I. P. Kushtalova, Recrystallization and Dispersion Hardening of Metals and Alloys [in Russian], Naukova Dumka, Kiev (1968).

335. G. V. Samsonov and S. A. Bozhko, Porosh. Met., No. 7, p. 30 (1969).

336. A. D. Le Claire, in: Diffusion in Body-Centered Metals (Proc. Intern. Conf., Gatlinburg, Tenn., 1964), American Society for Metals, Metals Park, Ohio (1965), p. 3.

337. S. D. Gertsriken and I. Ya. Dekhtyar, Diffusion in Solid Metals and Alloys [in Russian], Fizmatgiz, Moscow (1960).

338. G. V. Samsonov, in: High-Temperature Coatings [in Russian], Nauka, Moscow (1966), p. 7.

339. N. Engel, in: Diffusion in Body-Centered Metal (Proc. Intern. Conf., Gatlinburg, Tenn., 1964), American Society for Metals, Metals Park, Ohio (1965), p. 87.

340. J. F. Murdock, ibid., p. 262.

341. J. F. Murdock, ibid., p. 401 (discussion).

342. V. Z. Bugakov, Diffusion in Metals and Alloys [in Russian], Gostekhizdat, Moscow (1949).

343. I. Ya. Dekhtyar, Zh. Tekh. Fiz., 20:1015 (1950).

344. S. D. Gertsriken, Dokl. Akad. Nauk Ukr. SSR, No. 1, p. 72 (1956).

345. I. Ya. Dekhtyar, Dokl. Akad. Nauk SSSR, 85:583 (1952).

346. K. P. Turov, Dokl. Vyssh. Shkoly, No. 3, p. 152 (1958).

347. W. Seith and T. Heumann, Diffusion in Metallen: Platzwechselreaktionen, 2nd ed., Springer Verlag, Berlin (1955) [Diffusion in Metals: Exchange Reactions, US Atomic Energy Commission Report AEC-tr-4506 (1962)].

348. R. H. Moore, in: Diffusion in Body-Centered Metals (Proc. Intern. Conf., Gatlinburg, Tenn., 1964), American Society for Metals, Metals Park, Ohio (1965), p. 275.

349. G. V. Samsonov, Fiz.-Khim. Mekh. Mater., 4:502 (1968).

350. D. Graham, in: Diffusion in Body-Centered Metals (Proc. Intern. Conf., Gatlinburg, Tenn., 1964), American Society for Metals, Metals Park, Ohio (1965), p. 27.

351. D. Lazarus, ibid., p. 155.

352. R. L. Eager and D. B. Langmuir, Phys. Rev., 89:911 (1953).

353. P. L. Gruzin and V. I. Meshkov, Probl. Metalloved., Fiz. Met., No. 4, p. 570 (1955).

354. R. E. Pawel and T. S. Lundy, J. Phys. Chem. Solids, 26:937 (1965).

355. V. P. Vasil'ev, et al., Tr. Sredneaziat. Gos. Univ., No. 65, p. 47 (1955).

356. A. T. Tumakov and K. I. Portnoi (eds.), Refractory Materials in Machine Construction [in Russian], Mashinostroenie, Moscow (1967).

357. R. A. Swalin and A. Martin, J. Metals, 8:567 (1956).

358. V. D. Lyubimov, P. V. Gel'd, G. P. Shveikin, and Ya. E. Vel'mozhnyi, Izv. Akad. Nauk SSSR, Metally, No. 4, p. 132 (1966).

359. I. Ya. Dekhtyar, Dokl. Akad. Nauk SSSR, 85:583 (1952).

360. G. V. Samsonov and A. P. Épik, in: Investigations of Steels and Alloys [in Russian], Nauka, Moscow (1964).

361. G. L. Zhunkovskii and G. V. Samsonov, Fiz. Khim. Obrab. Mater., No. 4, p. 132 (1965).

362. G. V. Samsonov and G. L. Zhunkovskii, Porosh. Met., No. 6, p. 44 (1970).

363. V. I. Arkharov, in: Mechanisms of Interaction between Metals and Gases [in Russian], Nauka, Moscow (1964), p. 24.

364. B. Ya. Pines, Fiz. Tverd. Tela, 1:482 (1959).

365. G. V. Samsonov, I. F. Pryadko, B. M. Rud', V. V. Ogorodnikov, and L. F. Pryadko, Izv. Vyssh. Ucheb. Zaved., Fizika, No. 5, p. 120 (1970).

366. I. L. Sokol'skaya, Izv. Akad. Nauk SSSR, Ser. Fiz., 30:1996 (1966).

367. V. L. Bonch-Bruevich, Usp. Fiz. Nauk, 40:369 (1950).

368. A. A. Balandin, Usp. Khim., 33:549 (1964).

369. A. A. Balandin, Usp. Khim., 13:365 (1944).

370. M. Polanyi, Z. Elektrochem., 35:561 (1929).

371. W. M. H. Sachtler, Shokubai, 3:214 (1961).

372. N. I. Alekseev, Zh. Tekh. Fiz., 37:2224 (1967).

373. I. N. Frantsevich, R. F. Voitovich, and V. A. Lavrenko, High-Temperature Oxidation of Metals and Alloys [in Russian], Gosizdat, Tekh. Lit., Kiev (1963).

374. L. W. Swanson, R. W. Strayer, and L. E. Davis, Surface Sci., 9:165 (1968).

375. W. Gust, Report UCRL-6809 (1962).

376. G. V. Samsonov, Scientific Basis of Selection of Catalysts [in Russian], Nauka, Moscow (1966).

377. V. I. Makukha and B. M. Tsarev, Fiz. Tverd. Tela, 8:1417 (1966).
378. B. Ch. Dyubua, Radiotekh. Elektron., 10:1161 (1965).
379. G. V. Samsonov and A. I. Krasnov, Fiz.-Khim. Mekh. Mater., 2:485 (1966).
380. N. F. Lashko and S. V. Lashko-Avanesyan, Metallography of Welding [in Russian], Mashgiz, Moscow (1954).
381. N. F. Kazakov, Vacuum Diffusion Bonding of Metals, Alloys, and Nonmetallic Materials [in Russian], Izd. TIMMP, Moscow (1965).
382. B. V. Deryagin, What Is Friction? [in Russian], Izd. Akad. Nauk SSSR, Moscow (1952).
383. N. N. Zoren, Vestn. Moshinostr., 45(2):68 (1965).
384. A. Ya. Artamonov and V. I. Kononenko, Porosh. Met., No. 10, p. 84 (1966).
385. G. V. Bokuchaeva, in: High-Efficiency Grinding [in Russian], Izd. AN SSSR, Moscow (1962).
386. G. V. Samsonov, A. D. Panasyuk, and G. K. Kozina, Porosh. Met., No. 5, p. 36 (1968).
387. N. N. Rykalin, M. K. Shoroshov, and Yu. L. Krasulin, in: Nature and Chemical Bonding of Interstitial Metallic Phases [in Russian], A. A. Baikov Institute of Metallurgy, Academy of Sciences of the USSR, Moscow (1965), p. 37.
388. G. V. Samsonov, A. L. Burykina, and O. V. Evtushenko, Avtomat. Svarka, 19(10):30 (1966).
389. N. F. Kazakov, Abstract of Doctoral Thesis [in Russian], E. O. Paton Institute of Electric Welding, Academy of Sciences of the Ukrainian SSR, Kiev (1962).
390. G. I. Finch, Proc. Phys. Soc. London, B63:465 (1950).
391. G. I. Finch and R. T. Spurr, Brit. J. Appl. Phys., Suppl. 1, p. 79 (1951).
392. A. A. Shebzukhov and P. A. Savintsev, Izv. Vyssh. Ucheb. Zaved., Fizika, No. 10, p. 99 (1969).
393. A. P. Savitskii and L. K. Savitskaya, in: Surface Phenomena in Solutions and in Solid Phases Formed from Them [in Russian], Kabardino-Balkarskoe Knizhnoe Izdatel'stvo, Nal'chik (1965).
394. P. A. Savintsev and Kh. T. Shidov, Zh. Fiz. Khim., 40:1362 (1966).
395. P. A. Rebinder, E. D. Shchukin, E. V. Goryunov, and N. V. Pertsov, in: Abstracts of Papers presented at Fourth All-Union Conf. on Surface Phenomena in Melts and Solid Phases Formed from Them [in Russian], Kishinev (1968).
396. N. F. Lashko, S. V. Lashko, Soldering of Metals [in Russian], Mashinostroenie, Moscow (1967).
397. H. K. Adenstedt, J. R. Pequigntot, and J. M. Rayner, Trans. Amer. Soc. Materials, 44:990 (1952).
398. Yu. V. Efimov, Phase Diagrams of Metal Systems [in Russian], Nauka, Moscow (1968).
399. B. M. Askinazi, Hardening and Reduction of Parts by Electromechanical Treatment [in Russian], Mashinostroenie, Leningrad (1968).
400. A. V. Paustovskii, in: Abstracts of Papers presented at Fourth Conf. of Postgraduate Students and Young Scientists [in Russian], Scientific-Technical Information Division, Institute of Problems in Materials Science, Academy of Sciences of the Ukrainian SSR, Kiev (1970).
401. G. V. Samsonov and A. A. Zaporozhets, Fiz.-Khim. Mekh. Mater., 6(6):43 (1970).

402. G. V. Samsonov and Yu. G. Tkachenko, Porosh. Met., No. 3, p. 50 (1969).
403. G. V. Samsonov, Planseeber, Pulvermet., 15:3 (1967).
404. L. I. Struk, Abstract of Thesis for Candidate's Degree [in Russian], Institute of Problems in Materials Science, Academy of Sciences of the Ukrainian SSR, Kiev (1969).
405. O. A. Bulanin, Byull. TsILASh, No. 6, p. 53 (1937).
406. L. P. Tarasov, Machinist, 97:335 (1953).
407. G. V. Samsonov, A. Ya. Artamonov, and I. F. Idzon, Porosh. Met., No. 11, p. 75 (1967).
408. I. M. Fedorchenko and R. A. Andrievskii, Fundamentals of Powder Metallurgy [in Russian], Izd. AN UkrSSR, Kiev (1961).
409. I. Ya. Frenkel', Zh. Eksp. Teor. Fiz., 16:29 (1946).
410. B. Ya. Pines, Usp. Fiz. Nauk, 52:501 (1954).
411. G. V. Samsonov and V. I. Yakovlev, Porosh. Met., No. 7, p. 45 (1967).
412. G. V. Samsonov and V. I. Yakovlev, Porosh. Met., No. 8, p. 10 (1967).
413. G. Sh. Upadkhaya and G. V. Samsonov, Izv. Vyssh. Ucheb. Zaved., Tsvet., Met., 12(3):114 (1969).
414. I. I. Kornilov, P. M. Matveeva, L. P. Pryakhina, and R. S. Polyakova, Metallochemical Properties of Elements in the Periodic System [in Russian], Nauka, Moscow (1966).
415. Alloys of Rare-Earth Elements [Russian translation], Mir, Moscow (1965).
416. S. Z. Roginskii, Zh. Fiz. Khim., 5:175 (1934).
417. G. V. Samsonov, Kinet. Katal., 6:424 (1965).
418. O. V. Krylov, in: Scientific Basis of Selection of Catalysts [in Russian], Nauka, Moscow (1966).
419. J. E. Germain, Heterogeneous Catalysis [Russian translation], IL, Moscow (1961).
420. F. F. Vol'kenshtein, The Electronic Theory of Catalysis of Semiconductors, Pergamon Press, Oxford (1963).
421. Catalysis — Electronic Phenomena [Russian translation], IL, Moscow (1958).
422. S. Z. Roginskii, Zh. Fiz. Khim., 6:249 (1935).
423. L. D. Landau and E. M. Lifshitz, Electrodynamics of Continuous Media, Pergamon Press, Oxford (1960).
424. A. I. Brodskii, Physical Chemistry [in Russian], Goskhimizdat, Moscow (1948).
425. A. S. Davydov, Quantum Mechanics, Pergamon Press, Oxford (1965).
426. G. V. Samsonov, in: Scientific Basis of Selection of Catalysts [in Russian], Nauka, Moscow (1966).
427. A. Couper and D. D. Eley, Discuss. Faraday Soc., No. 8, p. 172 (1950).
428. G. V. Samsonov, Zh. Neorg. Khim., 8:1320 (1963).
429. G. V. Samsonov and N. N. Zhuravlev, Fiz. Metal. Metalloved., 1:564 (1955).
430. G. Rienäcker, Z. Elektrochem., 47:805 (1941).
431. G. Rienäcker, Z. Anorg. Chem., 264:54 (1951).
432. D. A. Dowden and P. W. Reynolds, Discuss. Faraday Soc., No. 8, p. 184 (1950).
433. A. A. Balandin, Usp. Khim., 31:1265 (1962).
434. H. S. Broadbent, L. H. Slaugh, and N. L. Jarvis, J. Amer. Chem. Soc., 76:1519 (1954).
435. G. A. Gaziev, O. V. Krylov, S. Z. Roginskii, G. V. Samsonov, E. A. Fokina, and M. I. Yanovskii, Dokl. Akad. Nauk SSSR, 140:863 (1961).

436. Liu Chung-Hui (Lyu Chzhun-Khuei), S. Z. Roginskii, and G. V. Samsonov, and M. I. Yanovskii, Neftekhimiya, 3:845 (1963).
437. V. M. Grosheva and G. V. Samsonov, Kinet. Katal., 7:892 (1966).
438. G. Duma and B. Galgoczy, Banyasz, Kut. Intez. Kozlemen., 6:221 (1961).
439. M. Merisalo, O. Inkinen, M. Jarvinen, and K. Kurki-Suonio, J. Phys. C., 2:1984 (1969).
440. M. I. Korsunskii, Ya. E. Genkin, V. G. Lifshits, and M. M. Omarov, Izv. Akad. Nauk Kaz. SSR, Ser. Fiz.-Mat., No. 2, p. 20 (1970); M. I. Korsunskii, Ya. E. Genkin, and V. I. Morkovnin, ibid., No. 2, p. 23 (1970).
441. F. M. Galperin and N. A. Gavrilov, Phys. Status Solidi, 40:53 (1970).
442. G. V. Samsonov, V. Ya. Naumenko, and L. N. Okhremchuk, Phys. Status Solidi a, 6:201 (1971).
443. I. R. Konenko, A. Nadzhm, A. A. Tolstopyatova, G. N. Makarenko, and G. V. Samsonov, Izv. Akad. Nauk SSSR, Ser. Khim., p. 2710 (1970).
444. I. R. Konenko, A. Nadzhm, A. A. Tolstopyatova, G. V. Samsonov, and G. N. Makarenko, Izv. Akad. Nauk SSSR, Ser. Khim., p. 100 (1971).
445. D. V. Sokol'skii and A. M. Sokol'skaya, Metals as Catalysts in Hydrogenation Reactions [in Russian], Nauka KAI, Alma-Ata (1970).
446. Z. A. Klimak and G. V. Samsonov, Kinet. Katal., 11:1394 (1970).
447. G. V. Samsonov and G. Sh. Upadkhaya, Porosh. Met., No. 2, p. 85 (1971).
448. N. N. Sirota, Physicochemical Nature of Variable-Composition Phases [in Russian], Nauka i Tekhnika, Minsk (1970).
449. G. V. Samsonov, M. S. Kovalchenko, R. Ya. Petrykina, and V. Ya. Naumenko, Porosh. Met., No. 9, p. 17 (1970).
450. M. S. Koval'chenko, R. Ya. Petrykina, and G. V. Samsonov, Porosh. Met., No. 9, p. 5 (1969).
451. N. N. Matyushenko, Crystal Structures of Binary Compounds [in Russian], Metallurgiya, Moscow (1969).
452. G. V. Samsonov and V. V. Morozov, Porosh. Met., No. 8, p. 55 (1970).
453. Thermal Properties of Solids [in Russian], Nauka, Moscow (1971).
454. Thermal Properties of Solids [in Russian], Naukova Dumka, Kiev (1971).
455. K. Schubert, Crystal Structures of Binary Phases [Russian translation], Metallurgiya, Moscow (1971).
456. Intermetallic Compounds [Russian translation], Mir, Moscow (1970).
457. Solid State Physics [in Russian], No. 1, Kharkov State University (1970).
458. E. A. Zhurakovskii, V. P. Dzeganovskii, and T. N. Bondarenko, Ukr. Fiz. Zh., 15:1476 (1970).
459. E. A. Bakulin, L. A. Balabanova, and M. M. Bredov, Fiz. Tverd. Tela, 12:72 (1970).
460. V. K. Grigorovich, Electron Structure and Thermodynamics of Iron Alloys [in Russian], Nauka, Moscow (1970).
461. M. M. Protod'yakonov, Properties and Electron Structure of Rock-Forming Minerals [in Russian], Nauka, Moscow (1969).
462. B. I. Markhasev, Izv. Akad. Nauk SSSR, Metally, No. 5, p. 170 (1970).
463. Thermionic Energy Conversion [in Russian], Atomizdat, Moscow (1971).
464. A. K. Mindyuk, Fiz.-Khim. Mekh. Mater., 6(6):60 (1970).
465. L. I. Gomozov, Izv. Akad. Nauk SSSR, Metally, No. 3, p. 146 (1969).

466. F. A. Cotton and G. Wilkinson, Advanced Inorganic Chemistry, 2nd ed., Wiley, New York (1966).

467. Physicochemical Properties of Oxides [in Russian], Metallurgiya, Moscow (1969) [The Oxide Handbook, IFI/Plenum, New York (1973).]

468. A. I. Voroninov, G. M. Gandel'man, and V. G. Podval'nyi, Usp. Fiz. Nauk, 100:193 (1970).

469. D. I. Khomskii, Fiz. Metal. Metalloved., 29:31 (1970).

470. B. A. Kovenskaya, Dokl. Akad. Nauk Ukr. SSR, No. 7, p. 661 (1970).

471. V. L. Ul'yanov and V. V. Nazarenko, in: Shrinkage in Alloys and Casts [in Russian], Naukova Dumka, Kiev (1970).

472. J. C. Slater, J. B. Mann, T. M. Wilson, and J. H. Wood, Phys. Rev., 184:672 (1969).

473. J. M. Ziman (ed.), The Physics of Metals, Vol. 1, Electrons, Cambridge University Press (1969).

474. S. M. Karal'nik, V. D. Dobrovol'skii, and A. V. Koval', "Electron structure of component atoms in transition-metal alloys and its relationship to crystal structure," in: Abstracts of Papers presented at Intern. Symposium on Electron Structure and Properties of Transition Metals, Their Alloys, and Compounds [in Russian], Kiev (1972).

475. M. I. Korsunskii, Ya. E. Genkin, and V. G. Zavodinskii, "Model of electron structure and properties of transition metals of the yttrium—palladium series," ibid., p. 127.

476. S. N. L'vov, P. I. Mal'ko, and V. F. Nemchenko, in: Metal Physics [in Russian], Kiev (1971), p. 22.

477. V. E. Zinov'ev, and P. V. Gel'd, Fiz. Tverd. Tela, 13:2261 (1971).